Assam Trucking Company

AIR TRANSPORT COMMAND,
BIRTH OF AMC

B. F. Bates

Franklin Publishing

PRINCETON, TEXAS

Copyright © 2020 by B. F. Bates.

All rights reserved. No part of this publication may be reproduced, distributed or transmitted in any form or by any means, including photocopying, recording, or other electronic or mechanical methods, without the prior written permission of the publisher, except in the case of brief quotations embodied in critical reviews and certain other noncommercial uses permitted by copyright law. For permission requests, write to the publisher, addressed "Attention: Permissions Coordinator," at the address below.

Kelly Carr / Franklin Publishing
1215 Juniper
Princeton, Texas 75407

www.FranklinPublishing.org

Ordering Information:

Quantity sales. Special discounts are available on quantity purchases by corporations, associations, and others. For details, contact the "Special Sales Department" at the address above.

Except where otherwise indicated, all Scripture quotations are taken from the New King James Version®. Copyright © 1982 by Thomas Nelson. Used by permission. All rights reserved.

Assam Trucking Company: Air Transport Command, Birth of AMC / B. F. Bates. —1st ed.

Copyright © 2020 by Franklin Publishing

Printed in the United States of America by Franklin Publishing, 2020

All rights reserved.
ISBN-10: 1-952064-99-6
ISBN-13: 978-1-952064-99-9

Testimonial

"Assam Trucking Company" greatly enriches the story of delivering cargo "Over the Hump." By chronicling the daily routines and operations and maintenance innovations, it highlights how much today's Air Force owes to the legacy of men and women who made great sacrifice and accomplished a herculean effort. Additionally, the knowledge gained about the planes, maintenance and logistics, as well as search and rescue and aeromedical evacuations were foundational. Without the lessons learned from this endeavor, the Berlin Airlift may not have had the profound strategic effect that it did, an effort regarded as airpowers' most decisive Cold War contribution.

CMSgt (Ret) Paul Wallace

Table of Contents

CHRONOLOGY 13
CHINA-BURMA-INDIA AIR TRANSPORT UNITS .. 18
TABLES 20
Acknowledgement 21
Chapter 1.0 - Prologue 25
  1.1   PROBLEM 27
  1.2   SOLUTION—A BELT AND SUSPENDERS. 33
      1.2.1   The Belt—Air Route 34
      1.2.2   The Suspenders—Ledo Road 35
  1.3   Let the Games Begin 35
Chapter 2.0 – Throwing Down The Gauntlet 38
  2.1   CHAOS & COMMAND 40
  2.2   TERRAIN 42
      2.2.1   India Side 43
      2.2.2   China Side 44
      2.2.3   Sabotage 46
      2.2.4   Chinese Chicken 47
  2.3   BURMA 48
      2.3.1   Hill Tribes 48
  2.4   WEATHER AND WEATHER REPORTING .. 49
      2.4.1   India Side 49
      2.4.2   Hump 52
      2.4.3   China Side 52
      2.4.4   Worst Season 52

2.4.5 Worst Day 53
2.5 WEATHER REPORTING 55
2.6 ALTITUDE 55
2.7 COMMUNICATIONS 56
2.8 NAVIGATION 58
2.9 JAPANESE 58
2.10 TONNAGE FIGURES 61
2.11 SHORTAGES 61
2.12 TURNOVER OF COMMANDERS 62
2.13 NUMBERS LOST BY ATC 64
2.15 AWARDS 64
Chapter 3.0 - Aviation Cadet Training 66
3.2 Operational Training 72
3.3 In-Theater Training 73
3.4 Jungle Indoctrination 76
3.5 Maintenance Training 78
Chapter 4.0 – AAFBUs 80
4.1 CONSTRUCTION 81
4.2 AAFBU ACTIVATION 84
   4.2.1 Mohanbari, 1332nd AAFBU 84
   4.2.2 Chabua, 1333rd AAFBU 86
   4.2.3 Misamari, 1328th AAFBU 86
   4.2.4 Sookerating, AAFBU 1337 & Tezpur, AAFBU 1327 89
4.3 LIVING QUARTERS 89
4.4 HEALTH & SANITATION 91
Chapter 5.0 – Morale 97
   Table 5-1, Rickenbacker Report Findings 99

5.1 DAY IN THE LIFE 101
  5.1.1 Ground Personnel 101
  5.1.2 Flight Crews 102
5.2 ENTERTAINMENT 105
  5.2.1 RADIO 106
5.3 ACTIVITIES 107
5.4 ROTATION POLICY 110
5.5 PERSONNEL 112
  5.5.1 Women 114
  5.5.2 Blacks 115
5.6 CRIME 116
5.7 PHYSICAL CHANGES 119
Chapter 6.0 – Planes 121
  6.1 C-47, Skytrain 122
    6.1.1 Technical Specifications 123
  6.2 C-46, CURTISS COMMANDO 123
    6.2.1 Technical Specifications: 129
  6.3 C-54, SKYMASTER 130
    6.3.1 Technical Specification: 131
    6.3.2 PERFORMANCE: 131
  6.4 B-24 CONVERSIONS 132
    6.4.1 C-87 132
      6.4.1.1 Technical Specifications: 132
    6.4.2 C-109 TANKER CONVERSION 133
ILLUSTRATIONS SECTION 135
Chapter 7.0 – Maintenance 142
  7.1 PRODUCTION LINE MAINTENANCE (PLM)
       145

- 7.2 PLM Stages 148
- 7.3 ENGINES 153
- Chapter 8.0 – Logistics 154
  - 8.1 SAFETY PROGRAM 155
  - 8.2 ATC UNITS 156
  - 8.3 OTHER UNITS 157
  - 8.4 CARGOS 157
    - 8.4.1 AVIATION GAS 158
    - 8.4.2 VEHICLES 158
    - 8.4.3 PX SUPPLIES 159
  - 8.5 REQUISITIONS 159
  - 8.6 CARGOES OUT OF CHINA 160
  - 8.7 PLANES 160
  - 8.8 PLANE PARTS 161
  - 8.9 SHORTAGES 162
  - 8.10 CHINESE TROOPS 163
  - 8.11 EFFECTS OF LOGISITICAL PLANNING .. 165
- Chapter 9.0 – Operations 166
  - 9.1 CREW SCHEDULING 169
  - 9.2 INCOMING DEBRIEF 171
  - 9.3 PRIORITIES AND TRAFFIC 174
  - 9.4 AIR CORPS SUPPLY 175
  - 9.5 OPERATIONS 175
    - 9.5.1 YOKE 175
    - 9.5.2 OPERATION ICHIGO 176
    - 9.5.3 RETREAT 177
    - 9.5.4 OPERATION GRUBWORM 177
    - 9.5.5 GALAHAD 178

9.5.6   ROOSTER MOVEMENT 179
Chapter 10.0 – Search & Rescue/Medevac 182
10.1   AAFBU 1352nd SEARCH AND RESCUE ... 182
10.2   AIRCRAFT 186
10.2.1   L-1 Stinson 186
10.2.2   L-4 Piper 187
10.2.3   L-5 Stinson 187
10.2.4   UC-64 Canadian Norseman 187
10.2.5   R-4 Sikorsky Helicopter 187
10.3   CRASHES & BAILOUTS 188
10.3.1   December 1943 189
10.3.2   March 1944 190
10.3.3   1945 191
10.3.4   JANUARY 192
10.3.5   February 1945 193
10.3.6   MARCH-JUNE 194
10.3.7   APRIL 197
10.3.8   JUNE 199
10.3.9   AUGUST 200
10.4   ACCIDENTS—January 1945 200
10.5   MEDEVAC 202
Chapter 11.0 – Closing Shop–1945 205
11.1   AAFBU 1329TH DERAGON 205
11.2   AAFBU 1328th, MISAMARI 206
11.3   AAFBU 1330th JORHAT 206
11.4   AAFBU 1327th, TEZPUR 207
11.5   AAFBU 1326th LALMANIR HAT 208
11.6   AAFBU 1333rd CHABUA 209

11.7 AAFBU 1332nd MOHANBARI 210
11.8 AAFBU KUNMING 212
Chapter 12.0 – What Happened To ATC? 214
  12.1 FEASIBILITY STUDIES ON THE FLY 214
  12.2 MILITARY AIRLIFT 218
    12.2.1 Military Air Transport Services (MATS), 1948-1966 218
    12.2.2 Military Airlift Command (MAC), 1966-1992 219
    12.2.3 Air Mobility Command (AMC), 1992 to Present 220
RETROSPECTIVE 221
  Appendices 225
  A   Pilot's Check List 226
  BEFORE STARTING ENGINES 226
  DURING WARM-UP 227
  BEFORE TAKE-OFF 227
  DURING FLIGHT 228
  BEFORE LANDING 230
  AFTER LANDING 230
  B   Glossary 232
  Acronyms 232
  DEFINITIONS 237
  C   End Notes 241
  Chapter 1—Prologue 241
  Chapter 2—The Gauntlet 243
  Chapter 3—Training 249
  Chapter 4—AAFBUs 250
  Chapter 5—Morale 251

Chapter 6—Planes 252
Chapter 7—Maintenance 253
Chapter 8—Logistics 254
Chapter 9—Operations 255
Chapter 10—Search and Rescue 256
Chapter 11—Closing Up Shop 257
Chapter 12—Where Did ATC Go? 257
D.   Bibliography 258
Works Cited 258
   Index 263
About the Author 275

CHRONOLOGY

YEAR	MONTH	EVENT
1931		Japan invades Manchuria
1933		Japan's invasion approaches the Great Wall of China
1937	May	Madam Chiang given control of Chinese Air Force
	June	Col. Claire Chennault arrives in China
	July	Marco Polo Bridge Incident
	Nov	Anti-Comintern Pact between Germany, Italy and Japan
1938		717-mile Burma Road built by hand
		Army Education Program
1939		CNAC began flying supplies to AVG
1940	Sept.	Tripartite Pact between Germany, Italy and Japan creating the Axis Powers
		US Army Officer Candidate School started in San Antonio
		USAAF AvCad Program

YEAR	MONTH	EVENT
1941		
	June	US Army Air Corps became US Army Air Force
	July	Japanese occupation of French Indonesia
	Aug	US Oil Embargo on Japan
	Dec	Bombing of Pearl Harbor
		Bombing of Singapore
1942		
	March 8	Rangoon captured by Japanese
	March	Activated Chabua
		Mohanbari
		Tezpur (British Base)
	April	Regular airlift with CNAC and other groups began
		ATC established
	July 8	Pub Law 658, First Officer Act
	July 28	First C-47 landed at Chabua
	Sept	First night flight by Capt. John D. Payne
		First Communication unit moved to Chabua
	Oct 25	Limited Allied air transport

YEAR	MONTH	EVENT
		First air raids at Chabua and Mohanbari
	11 Nov	621st Air Evacuation Squadron activated at Bowman Field KY
	Dec	Ledo Rd. construction began
		Organizational control of ATC moved to Washington, D.C.
	Dec 25	First 15 C-87s arrived at Chabua
1943	July	2,916 net tons airlifted to China
	Sept	Project 8
	Oct	Capt. John L. "Blackie" Porter officially began Search and Rescue unit at Chabua
		Porter named Flying Safety and Rescue Officer
	Nov	Jungle Indoctrination Camp
	Dec 10	Porter died during a rescue
		Simulator arrived at Chabua
		Misamari operational
1944	Feb	1352nd Search and Rescue moved to Mohanbari

YEAR	MONTH	EVENT
	March	American Army Railway Operating Battalion took over running the Bengal-Assam Railway
	April	YOKE Operation
		Sookerating in operation
	July	18,975 net tons airlifted to China
	August	Myitkynia recaptured
	Dec	Operation Grubworm
1945	Jan	Worst weather experienced over the Hump
		First convoy left Ledo for China
		Scheduled passenger and freight flights started
	Feb	PLM started at Jorhat
	March	Link Trainer received at Misamari
	March 15	Pvt Perry executed after being convicted of murder
	April	Butler hangar in use at Misamari
		Operation Rooster Movement

YEAR	MONTH	EVENT
	May	Butler hangar in use at Mohanbari
	May 27	16th AFR Station VU2ZK began broadcasting
	29	Last Scheduled flight to China from Tezpur
	31	Tezpur deactivated
	June	Deragon closed as an ICD base
		Jorhat received an ice plant
	July	Declassification of APO
		71, 042 net tons airlifted to China
	Sep 19	Orders for deactivation received
	20	Last C-87 with a load for China flew out of Jorhat
	26	Last Operational C-87 flown to Bangalore
		Hope Project began taking passengers from Chabua to Calcutta
	Oct	Deactivation of Chabua and Mohanbari started end of month
	11	Jorhat deactivated

CHINA-BURMA-INDIA AIR TRANSPORT UNITS

Army Air Force Base Unit	Location	Code Sign
1304th AAFBU	Barrackpore	CM
1305th AAFBU	Dum Dum	
1306th AAFBU	Karachi	
1311th AAFBU	Gaya	
1326th AAFBU	Lal Hat	
1327th AAFBU	Tezpur	YP
1328th AAFBU	Misamari	
1329th AAFBU	Deragon	
1330th AAFBU	Jorhat	PW
1332nd AAFBU	Mohanbari	KC
1333rd AAFBU	Chabua	VG
1337th AAFBU	Sookerating	GH
1339th AAFBU	Chengkung	DB
1340th AAFBU	Kunming	RQ
1342nd AAFBU	Chanyi	CY
1343rd AAFBU	Luliang	IM

1345th AAFBU	Kurmitola	GI
1346th AAFBU	Tezgaon	CV
1347th AAFBU	Shamshernagar	BQ
1348th AAFBU	Myitkyina	FC
1352nd AAFBU	Search & Rescue	
1359th AAFBU	Loping, China	

TABLES

Chapter	Graphic	Table #	TABLE
1	Prologue Time Line		
2		2-1	ATC Hump Command & Tonnage
		2-2	ATC Aircraft Losses
3	Time Line		
5			
		5-1	Capt. Rickenbacker's Observations
		5-2	Courts Martial-- Misamari
9		9-1	Tonnage Delivered by ICD ATC
10		10-1	January 1045
		10-2	Breakdown of Accidents by Base unit
		10-3	Breakdown by Cause

Acknowledgement

Research for a graduate history paper on the Air Transport Command in China-Burma-India began in the Austin College library with a review of available sources, most of which, with a few exceptions, were third-person accounts. The review revealed that since the end of World War II, many volumes about various military units, battles, and strategies have been written. Little, however, had been written about the Air Transport Command (ATC)—its mission, men, and planes—in the China-Burma India (CBI) Theater. Of the official military histories available, only Craven and Cate's series, The Army Air Forces in World War II, Volume 6, Men and Planes, and Volume 7, Services Around the World, and Romanus and Sunderland's series, U.S. Army in World War II, Volume 9, Part 2, China-Burma-India Theater, Stilwell's Command Problems, and Part 3, Time Runs Out in CBI, discuss the Air Transport Command's mission in any detail.

In addition to the official histories, there are several other sources for information including Barbara W. Tuchman's Stilwell and the American Experience in China, 1911-1945; a four-volume series titled, China Airlift—The Hump, produced by and for the Hump Pilot's Association; Thunder Out of China by Theodore White and Analee Jacoby; and, Ten Thousand Tons by Christmas by Edwin Lee White.

Aside from the previously mentioned books and resources, I found the remembrances of General William H. Tunner, the last commander of the Air Transport Command (ATC) in China-Burma-India (CBI), entitled Over the Hump, to be a good source. In spite of the title, Gen. Tunner discusses the Hump flight operations from his assumption of command in August 1944 and his subsequent career in the

Military Air Transport Service (MATS) and as commander of the Berlin Airlift. His family has given their permission to use information from his book.

When I was encouraged to expand the original paper, new research was required. The Air Mobility Command, Historian's Office at Scott AFB, IL, proved to be a treasure trove of materials encompassing the most interesting primary historical collection of monthly reports by Army Air Force Base Unit (AAFBU) historians from each AAFBU through three years of the Hump operations. In addition, access to US Army Air Force personnel military records, personal letters, in-theater newspapers, the Hump Pilot's Association works in three volumes plus monthly newsletters and interviews with some of the men who flew the Hump rounded out the first-person sources as the basis for the following work.

While other units were involved in the China-Burma-India Theater, the focus of this particular book is the institution of the first military airlift in the history of the U.S Army Air Corps/Air Force and what developed from lessons learned to address changing global requirements, resulted in Air Search and Rescue, Medical Airlift capability, Production Line Maintenance (PLM), and logistical airlift on a greater scale than ever could have been perceived. What was developed and refined over the Hump, was put into use just two years later as a literal lifeline for the people of Berlin by the last commanding general of the Hump operations, Gen. William Tunner.

During the research phase, trips to the U. S. Air Academy in Colorado Springs and the Air Mobility Command's Historian's office at Scott AFB, IL, opened a number of primary sources to me, including monographs, unit histories, in-theater newspapers, orders-of-the-day listing citations, awards, and promotions, etc., and letters.

In later years at Hump Pilot Association reunions, I conducted interviews with others who had experiences in the CBI Theater. The following men provided their memories of life on the Hump:
- John M. Foster, Mohanbari, 1944-45
- Gordon Smith,
- Lonnie Johnson, Misamari, 1945
- Paul Schaffer, Kurmitola, 1944-1945
- Gordon J. Leonard, Jorhat, 1944-1945
- William McCoy, Jorhat, 1943-1944
- Robert M. "Pete" Loving, Jr., Chabua 1944-1945

The Hump Pilot's Association gave me permission to use their materials from the multi-volume set, *China Hump Pilots*. The Association has since closed its doors due to a shrinking population of the men and women who flew the Hump.

My family has been very supportive of my research through the years. In addition to those already noted, Mrs. John M. Foster has been a major support and advocate for this effort to the point of bequeathing me my father's military records and flight charts from this period of his service. Col. Thomas R. Foster, USAF (ret). has also provided his expertise and support. Ms. Beverly Mardis and Dr. Don Hawkins have given their support as advisors in the writing effort, with editing and critiques throughout the process. Most important of all were the men who left the comfort of their homes in the United States to fight in a foreign country. To a man, I did not hear a negative word, as we did during the Vietnam war. When asked if they questioned why they were being sent to an unknown area of the world to fight people with whom they had no personal complaint, their response was that they were fighting to keep the Japanese from joining up with the Germans and the Italians. They had a responsibility to their country. Their stories are included in the following pages.

Chapter 1.0 - Prologue

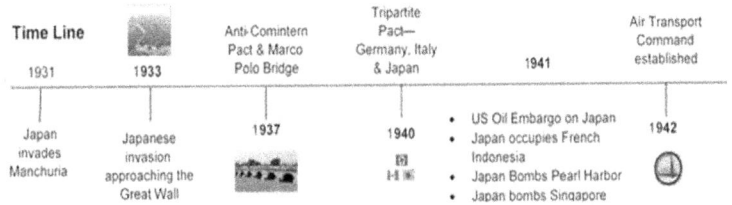

Developed and proven in World War II in the China-Burma-India (CBI) theater over the Himalaya Mountains (known as the Hump), a fledgling airlift effort evolved into the Tactical Airlift capability employed by the Air Mobility Command today. Considered one of the most dangerous air routes in the world, with tumultuous weather and ever-present, marauding Japanese fighters, General William H. Tunner, last commanding general of Hump operations, said "It was safer to take a bomber deep into Germany than to fly a transport plane over the 'Rockpile' from one friendly country to another."[1] The amazing fact about the effort is that it literally "got off the ground," and became an effective mode of tactical airlift of logistical supply.

Why an airlift? Why and where was it needed? When the call for support to China was first issued, the United States was sitting on the sidelines and watching. As early as 1933, the United States leadership, aware of the various factions and the struggle going on in China, discussed the ultimate possibility of the United States entering the war with Japan, yet did nothing. Between July 7, 1937, and December 1, 1941, the United States played a waiting game regarding the world situation, not wanting to be drawn into the war as an active participant despite the December 1937 bombing of the USS Panay gunboat by the Japanese in the Yangtze River.[2] The U. S., still a supporter of Japan, was growing more suspicious of their intentions. Unpopular with the U. S. Congress, trade suppression and embargo were the only diplomatic options being considered to influence Japan. Not wanting to be a party to war, isolationist policies held in Congress were the guiding force even though the world as a whole appeared to be careening inextricably toward world war.

The U.S.'s position changed on December 7, 1941, with the bombing of Pearl Harbor. No longer on the outside looking in, the United States had been brought into the war, like it or not.

Due to the war's massive scale, World War II presented a number of challenges. New aircraft weapon systems became major players in the ability to move men and supplies to remote areas of the world. The Ferrying Command, activated as one entity with the mandate to deliver Lend-Lease aircraft to Great Britain, became two commands in June 1942, to address the ever-increasing logistical supply requirements of war. The Ferrying Command continued delivery of aircraft to Great Britain while the newly formed Air Transport Command (ATC) became the delivery vehicle for supplies around the world.[3] Specifically, ATC would be used in the airlift effort into China from India.

1.1 PROBLEM

After the Japanese invasion of Manchuria in 1931, National Chinese leader, Generalissimo Chiang Kai-shek, found himself literally caught between the Communist Chinese faction under Mao Zedong in northwestern China and Japan in the east. From 1931 until July 1937 the relationship between the two countries was strained. China, always in need of financial help, asked for aid from Italy, Russia and the United States to support its war efforts, primarily against the Communist Chinese and not Japan.

In July 1937, a confrontation between a contingency of Japanese soldiers on training exercises and a group of Chinese soldiers occurred near Beijing at the Marco Polo Bridge. After the confrontation, a Japanese soldier, Private Shumira Kikujiro, was reported missing. Seizing the opportunity to further their aggression, the Japanese demanded that they be allowed to search the nearby walled town of Wanjing for the missing soldier. Chinese commander Ji Xingwen refused their request. Reinforcements for both countries arrived as tempers intensified. As both sides continued mobilizing along the Yondong River, Private Kikujiro, who had become lost when he wandered away to relieve himself, returned unharmed. Despite his return, a Japanese infantry unit tried to breach Wanjing city walls but was turned back. The Chinese went on heightened alert. The Japanese opened fire, but the Chinese held the bridge.[4]

Japanese military members and Foreign Service representatives began negotiations with the Chinese National Government. A verbal agreement was reached, and Japan apologized to China.

Japan, however, took advantage of the incident to invoke their right to be in China under the Boxer Protocol of 1901.[5]

Japan had, in fact, up to 12,000 soldiers already quartered in Beijing under the guise of protecting its diplomatic mission—considerably more than was needed. From that point on, Japan steadily took control of China with little effort. Chiang's focus remained on Mao and the Communist Chinese as Japan continued to grab Chinese territory,

In May 1937 the Generalissimo gave his wife, Madam Chiang, control of the Chinese Air Force because she was the only person he could trust. Her advisor, U. S. Army Air Corps (AAC) pilot Roy Holbrook, recommended she hire former U.S. Army Air Corps Captain Claire Lee Chennault to lead the fledgling Chinese Air Force.[6] Chennault, having left the Army Air Corps in April 1937 due to health issues and disputes with his superiors, was well known as a troublemaker and viewed as a mercenary by U. S. Army Generals George C. Marshall (Commander-in-Chief of the U. S. Army in WWII) and Hap Arnold (U. S. Army Air Force Commander). His bad reputation with his former Army colleagues was made worse as he became involved with Chiang.[7] Nevertheless, the now self-appointed Colonel Chennault arrived in China in June 1937 and began his duties. As the Japanese invasion spread to Tientsin and Shanghai, he rallied the Chinese pilots, most of whom did not come back alive. Under Chennault's command, the mercenary American Volunteer Group (AVG), better known as the "Flying Tigers," was formed. It was their job to keep the Japanese at bay while giving support to the Chinese forces.[8]

Madam Chiang and her brother, T. V. Soong, educated in the States and well-versed in the workings of the U.S. government, became the voice of the Chinese regime in the United States. Soong, Chiang's emissary in Washington, D. C., had the ear of Averell Harriman, director of the Lend-Lease program in charge of funding and war matériel.

Assam Trucking Company

By 1938 the Japanese had driven Chiang further westward into the mountainous Szechuan province. While they retreated, the Nationalist Chinese seized capital from private and public sources to keep their effort alive. In the summer of 1939, Chiang needing more support, made another appeal to FDR for assistance using the Japanese threat as justification.[9] A propaganda campaign, opened in the U. S. by Madam Chiang and her brother, blamed the West for letting China down. W. H. Donald, an Australian newspaperman and advisor to Chiang, who had written about Japanese aggression, wrote, "if the Democratic powers refuse to give China aid, they will commit the greatest crime in history, if they, in any way, succor or give tangible support to the Japanese."[10] Because of Donald's advice and warnings against them, the Japanese offered a reward for his capture—dead or alive.

Japanese expansionism was on the move as it continued its mission of aggression on the Chinese mainland. By 1940, Japan lost no time in closing off the entire Chinese coastline of 18,000 km. As the situation in China shifted with more territory occupied by the Japanese, China's predicament became unmistakable.

At the Imperial Japanese conference of July 21, 1941, the Japanese made a decision to move into Indochina [Vietnam, Cambodia, Laos, Thailand, W. Malaysia, and Burma (Myanmar)] and to enter into a tripartite agreement with Germany and Italy to keep the British and U. S. off-balance in the Pacific Theater as Germany and Italy conquered Europe and moved into North Africa. Most of the men sent to China-Burma-India (CBI) felt they were keeping the Axis Powers from joining forces to overcome the world.

"Japan was trying, on order of the Axis, to link up with Germany and Italy. The push was being made into China, splitting it in two. Another push came through Malaya and Burma toward India as far as the Assam Valley."[11]

Since the U.S. held no territory in Asia, they saw no reason to do anything more than throw financial aide at the Chinese to appease them, while the *real* war was fought elsewhere. Available military units were needed in the European and Pacific Theaters. The solution perceived—providing supplies and munitions through Lend-Lease—required minimal air support, technicians and training without a large commitment of personnel. All things considered, the Chinese had an unlimited supply of manpower, so the Allies felt the solution would have no impact on the war elsewhere. The "primary purpose of the Allied effort in CBI was *not* to win the war but to hold the enemy in check while victory was achieved elsewhere."[12] The powers that be felt it was more important to keep the Chinese engaged in case they were needed later in the war as a base of operations for an assault on Japan.

As the Japanese advanced into Burma, the huge backlog of provisions in Rangoon and Lashio, the terminus for Burma's primitive railroad, was in jeopardy of being captured. To add to the problem, all the rail lines and roads ran south to north through the Salween, Irrawady, and Mekong river valleys between the three separate mountain ranges—Patkai, Kumen, and Santung—of the eastern Himalayas. There were only a few trails transecting the road system in the rugged mountain barrier separating India from Burma. A trickle of Lend-Lease goods and other outside aid was shipped to Lashio, the northern terminus of the railroad. At Lashio the supplies were transferred to trucks for transport over the Burma Road, a torturous 717-mile long dirt track built by hand in 1938.[13] During the monsoon season it was barely-passable, a mud trail from Myitkyina (Mish-i-naw), Burma, to Kunming, China. Until its capture by the Japanese on March 8, 1942, Rangoon had served as the main port-of-entry for supplies going to China—its *only* land link with the outside world. With its fall, a new route and port-of-entry needed to be quickly identified.

Now the problem of providing logistical support and supplies to the Chinese became one of coordination of efforts and transportation. Everything needed by the Chinese from paper clips and airplane parts to aviation fuel, had to be transported through the impenetrable jungles of Burma to the Burma Road.

China's bargaining position with the U. S. and Great Britain changed dramatically with the bombing of Pearl Harbor and the entry of the U. S. into the war. The Chinese now felt they could put the U. S. on the spot.

With the capture of Rangoon and the new port-of-entry for CBI identified as Calcutta, India, the movement of the provisions to troops in China became the issue. The Japanese invasion and control of the Chinese mainland seaports and Burma, and the terrain and primitive transportation available within India and Burma, required an unconventional approach. Adding to the problems were the differences in "national interests of Great Britain and the United States. Almost two years of the war passed before a concrete program had been agreed upon."[14] The three commanders in the theater, each with their own agenda to support, were from three different countries and all in need of supplies.

- Chiang wanted control of *all* U. S. and British assets to address his hidden agenda of hoarding military supplies for his war against the Chinese Communists, once the Japanese issue was taken care of by the U. S. and Britain.
- U. S. Commander General Joseph Stilwell sought to have control over the Hump operations including the China side.

- Lord Louis Mountbatten, British Supreme Allied Commander, South East Asia Command, was trying to keep the Japanese out of India and protect the Crown's interests in the area.

"The India-Burma Theater mission was further complicated by the unusual overlapping of British and American geographical areas of responsibility and the difference in national objectives in Southeast Asia and China."[15] Once the mission of carrying supplies to China was established, it would be expanded to include ferrying supplies to various Chinese and American bases of operation within the interior.

"American Command was designated as China-Burma-India (CBI)"[16] covering the three countries indicated. The primary interest in each of the three commands was as follows:
- American mission to use India as a:
  - Base for the establishment of air and land routes from the Assam Valley to the Yunnan Province in China
  - Springboard for forwarding supplies to the forces fighting the Japanese in China
  - Tactics to keep the Chinese bases used by the B-29 crews for bombing runs on the Japanese mainland from being captured by the Japanese
- The British to re-conquest of:
  - Burma
  - Federated Malay States and Singapore
  - Andaman and Nicobar Islands
  - Sumatra
  - Thailand [17]
- Chinese to keep the supply train moving to Chiang

As leader of the Kuomintang (KMT), the ruling party of China, Chiang had other personal concerns in local corruption, blackmail, black market, embezzling, extortion and general terrorism over the common Chinese.[18] He was nominally in the war, with no real commitment other than to prop up his own agenda.

1.2 SOLUTION—A BELT AND SUSPENDERS

With the Japanese advancing through Burma, it became apparent that the Burma Road would no longer be available. Shortages of naval assets would never permit a seaborne invasion of the Burma coast. The Allied re-conquest effort of this strategically important country would have to be based in India to the west, over lines of attack dominated by the towering Naga and Chin Hills. The "hills" were actually mountains with peaks up to 10,000 feet.[19] To expedite aid to China, Soong showed Harriman a map of southwest Asia displaying an alternate land route. This alternate route, covering a total of 5,000 miles, would utilize the Persian Gulf in the Soviet-held territory of Iran. The route included 840 miles of railroad from the Persian Gulf to the Caspian Sea, where the cargo would be shipped by boat to the Russian Turkestan Railroad then for 2,000 rail miles to Sergiopol near the Chinese border. Finally, the cargo would be transported by road 2,000 miles to Chungking.[20] Besides the overland route through Soviet territory, several other routes were proposed among them were:
- A new road to be built from Ledo, India, to a point on the Burma Road north of Japanese occupation
- A pipeline to parallel the new road
- An air route

1.2.1　The Belt—Air Route

Chiang and Chennault both felt the air route from India to China would be more effective than the proposed road. At the time he was briefed on the proposed air route, General Stilwell agreed that the air route to China was needed no matter what happened to the Burma Road,[21] which was soon inaccessible because of Japanese occupation. The new proposed road would be constructed farther north from Ledo, Assam, India, joining with the portion of the Burma Road not under Japanese control.

The China National Aviation Corporation (CNAC), established by Pan American World Airways to fly commercially in China and owned co-jointly by the Chinese (55%) and Pan Am (45%), had been in operation several years before the start of the war. [22] In 1939 CNAC began flying supplies and personnel to isolated areas within China. Most of Chennault's supplies for the AVG in 1941 and 1942 arrived by CNAC planes flown from Calcutta to Kunming. This was the route Soong referred to in a letter to President Roosevelt. What Soong failed to mention was a "small obstacle" existed in the form of the Himalaya Mountains (the Hump). Within nine days President Roosevelt, upon the recommendation of Harriman, approved the airlift operation. Harriman had been in China and felt that it was to the benefit of the Allied forces to keep China in the war.[23]

The port-of-entry for provisions to China would be Calcutta on the northeastern coast of India. It was estimated that 100 DC-3s, once in theater (75 for U. S. Army Air Force and 25 for CNAC), could move 5,000 tons per month.[24] Allied air transport in July 1942 was extremely limited. The Royal Air Force Squadron 31 had a few Dakota (the British variation of the C-47) transports, which provided a very small amount

Assam Trucking Company

of supply and evacuation support to the combined British, American and Chinese forces.[25]

Because CNAC had been flying the suggested route on a commercial basis, it was felt the newly organized Air Transport Command (ATC), could handle the task at hand. The air route to the non-existent Indian terminals and then to Kunming was characterized as "comparatively level stretches of 550 and 700 miles respectively."[26] The route was actually one of the worst in the world, with the comparatively level stretches at the beginning in India and at the end of the route in China. It was at Soong's insistence that an air bridge was seen as a more viable supply route that led to U. S. involvement in the logistical challenge that became the Hump.[27]

The tortuous course through Soviet-held territory was not used in deference to the other shorter three methods, each of which used the same general route from India.

1.2.2   The Suspenders—Ledo Road

The "new" track to connect to the Burma Road north of the Japanese occupied territory would be about 500 miles long. It had taken two years to build the Burma Road. The new road would take just as long to construct if the Japanese push into India was stopped. Begun in December 1942, the Ledo Road was not completed for two years—two years Chiang did not have. His need of an estimated 100,000 tons per month was immediate.

1.3 Let the Games Begin

The air road started operations in the spring of 1942 with CNAC and the varying commands. As with any other multinational venture, there was jockeying for position and control. CNAC felt that because they had the experience along the route, they should be in charge of all Hump airlift operations, and the military should "operate under the airline's general direction."[28]

General Joseph Stilwell "who was more interested than any other American in getting supplies over the Hump, answered in emphatic terms. He said it would be unfair and bad practice to put civilians, who were not subject to military discipline, over Army pilots, who were subjected to greater dangers for much less pay. Adequately regular operations by the civilian airline could not be assured unless it was placed under military control. Without such control, it was impossible to keep the airline from being forced by Chinese officials to waste priceless capacity flying in goods for profitable resale. It was similarly impossible to keep the planes from being diverted to highly-paid missions inside China." Chiang Kai-shek concurred, and CNAC was placed under U.S. government contract. [29]

"Air Transport Command (ATC) was to supply the Chinese Nationalists in southern China in their fight against the Japanese. This became necessary because of the Japanese occupation of all key ports in China. The subsequent capture of the only land route, the Burma Road, across the southern end of the Himalayas made it necessary to construct another road—the Ledo Road (Stilwell Road)—across the Himalayas farther to the north through Myitkyina and to fly over the 'Hump' with food, fuel, munitions and troops."[30] Gen. Stilwell felt that the road, once built, would be far more effective than an airlift. He accepted the estimates that over 100,000 tons of supplies could be carried over the road at its peak performance. Furthermore, Stilwell had General Marshall's backing.[31] In reality the Ledo Road was never a big player in the supply chain. In its best month only 6,000 tons of supplies

Assam Trucking Company

were delivered over the Ledo Road while the flights eventually were able to deliver 71,000 tons in a month.

ATC in China-Burma-India (CBI) was stretched to the limit at the end of a 12,000-mile-long, supply line with headquarters in Washington, D.C. ATC's customers considered themselves all Priority Number One and included:
- Chinese Air Force, to become the 14th Air Force under Gen. Claire Chennault
- British functions of General Orde Wingate's Chindits in Burma
- U. S. Army undertakings in China and India
- Merrill's Marauders in Burma 32

As the theater came into full operation, the original mandate of supply was expanded to incorporate medical and civilian evacuations, troop movement operations, search and rescue of downed crews, and movement of personnel and supplies within China. Unlike the European and Pacific Theaters in WWII who received their supplies by ship convoys, the units in China were cut off from any supply by ship or land. The following stories chronicle the birth and growing pains of an airlift effort over the Hump into China— the first of its magnitude and kind in military history.

Chapter 2.0 — Throwing Down The Gauntlet

At the beginning of Hump Operations in 1942, the battle cry of airlifting ten thousand tons by Christmas seemed doable. At first all parties were in accord; however, Murphy's Law prevailed—whatever could go wrong, did. It turned out to be a logistical nightmare. There was no blueprint, no strategy, no plan, no precedence, not even a plan scribbled on a napkin. Where to start? For ATC the center of command was in Washington, D.C. There was no in-theater leadership. The challenges seemed insurmountable; but the nature of the mission was, in and of itself, the greatest challenge faced by the men of ATC as it was the

"Sole means by which a combat theater was nourished. Every vehicle, every gallon of gas, every weapon, every round of ammunition, every typewriter, and every ream of paper which found its way to Free China for either the Chinese or the American forces during nearly three years of war was flown in by air from India." [1]

The operation was made more difficult because India was not the primary source of supply. The fact that most supplies had to be shipped into one of three Indian ports—Calcutta (Kokata), Bombay (Mumbai), or Karachi (Pakistan)—by sea or air added to the nightmare. Once in country, the supplies had to be moved to an Army Air Force Base Unit (AAFBU) in the Upper Assam Valley in far northeastern India. From Calcutta the transportation system attending the tea plantations and deemed adequate for military supply, subsisted as a complicated transportation system of primitive, varying-gauge railways with limited carrying capacity and capabilities in conjunction with slow river barges on the Brahmaputra River. Because the gauge of the rails between systems was not standardized, cargo had to be unloaded from one train and reloaded to another train of the gauge needed. And because no direct roads or bridges existed, the rail cars had to be placed on river barges at points along the way to cross the Brahmaputra River before continuing the trip. To add to the hurdles, Calcutta, in southeastern India the closest of the three ports, was harassed by Japanese bombers and fighters. [2]

Before the base facilities were even constructed, creation of the groups to be assigned to the ATC in CBI began. Activated as the Third Ferrying Squadron, First Ferrying Group from Westover Field, MA, the 1333 AAFBU, Chabua, Assam, with the 1332nd from Pope Field, Ft. Bragg, NC, shipped out of Charleston, SC, on March 17, 1942, at 2100 hours on the USA Transport Brazil, destination unknown. After clearing Port Elizabeth, South Africa, all doubt disappeared about their final port-of-call. With the education programs and briefings covering such topics as "Language of India," "Jungle Warfare," and "Customs of the Hindus," the men knew they would be part of the effort based in India. Upon arrival at Karachi on May 17th, after 58 days at sea, they found the facilities there sorely lacking, that is to say—

nonexistent. [3] No administrative personnel met them for in-processing or to arrange billeting. The new arrivals set about almost immediately building rudimentary facilities. The men recycled P-40 packing crates to build an orderly room (3 crates), a kitchen (4 crates), and a mess hall (6 crates). [4]

On August 1, 1942, the Third Ferrying Squadron boarded the train for Chabua. Army personnel from diverse walks of life in the States were treated to a view of a very different part of the world, far removed from their personal experiences.

"India's trains were quite like our comic strip 'Toonerville Trolley'. The men were given an opportunity to view India that was, as yet, a land of charming mystery to them. Mile after mile of flat, desert-like plain, the surface one of magnificent failure, was viewed. Areas flooded by recent rains, rice paddies, water buffalo mingling with rough coated, half-starved, mangy, long-eared, camels, and wide-eyed natives, all passed in review like characters in a Disney Technicolor production. The troops, despite discomfort of the crowded trains, numerous flies and mosquitoes, were seeing India — the land of varied scenery and lavish color; the land of extremes of riches and poverty; a land of heat and cold, disease and dirt; a fascinating land of beauty and mystery, of slavery and freedom dimly sensed but out of reach." [5] The trip of 1,000 miles took three weeks, with frequent stops for changes in train equipment and re-routings, and to address security concerns caused by bandits and members of political parties opposed to the British.

2.1 CHAOS & COMMAND

Another element of the airlift equation rested in two formidable personalities each with a clear sense of leadership, in the persons of Generalissimo Chiang and Lord

Assam Trucking Company

Mountbatten. They clashed, much like the weather systems crossing the Hump, each feeling they should receive *all* supplies delivered to CBI. Because they could not play well together, the theater was soon reduced to chaos, with no apparent central operational command. Army Chief of Staff General George C. Marshall stepped into the fray and appointed General Joseph Stilwell, who was on the same command level as Eisenhower in Europe and MacArthur in the Pacific, to lead the initiative.

With his work cut out for him as a referee, Stilwell was the perfect person for the job, having spent three tours of duty in China after WW I. He had an understanding and appreciation for the Chinese people and the language that others in the command did not have. Despite that, he expressed a defeatist attitude, saying he "felt it was fantastic to think such an operation could work." [7] No matter how he felt, the airlift was going to happen.

Even with Stilwell in command, operations were slow to start. A shortage of aircrews resulted in flights not being scheduled 24/7 as originally planned. Only three bases were under construction in the Upper Assam Valley during the winter of 1942-43—Sookerating (Sook-er-ting), Chabua (Cha-bwa), and Mohanbari (Moh-an-berri). By the beginning of the monsoon season in March 1943, only Chabua, which became the main point-of-entry and the embarkation terminal for the duration of the war, was fully operational. The other two were closed due to lack of adequate runways, taxiways, and hard standings. The British engineers in charge of construction at Sookerating and Mohanbari had failed to do their jobs as witnessed when an *empty* transport taxied down the turf runway, leaving deep ruts in the surface. Because of the condition of the runways, there was no flying done from Mohanbari" [8] until late 1943.

The early problems of an antiquated or primitive communication system, no accurate weather reporting system and constant diversion of flights to move refugees or troops away from advancing Japanese, were just several factors confronting Stilwell as commander of the CBI Theater.

Adding to the command problems and the conditions of the base units under construction, the challenges facing the ATC aircrews in their mission to deliver the necessary provisions needed included the:

- Terrain
- Weather and weather reporting system
- Altitude
- Communications
- Navigation
- Japanese
- China Terminals
- Shortages
- Commander Turnovers

2.2 TERRAIN

Lt. Tom Harmon, P-38 fighter pilot with the 14th AAF, stated in 1943, "I would rather fly a combat mission any time than tangle with that flight over the Himalayas." [9]

The terrain found in the Theater varied from thick jungle growth to barren mountain peaks which resembled rock piles above the tree line. Within ten minutes of takeoff from the Assam Valley aerodromes, any sign of civilization disappeared under the thick, green jungle canopy. Recognition of landmarks, aside from the major rivers and mountain ranges of Southeast Asia, was impossible. [10]

2.2.1 India Side

Flying east from Assam, the transports climbed steadily at the rate of 300 feet per minute at a speed of 145 mph for the first 125 miles. Altitude at the Assam bases was less than 100 feet above sea level. The Chindwin River Valley came next, then the Kumen Mountains at 14,000 to 16,000 feet. These mountains separated the west and east branches of the Irrawaddy River. Between the next two great river gorges, the Salween and the Mekong, was the main Hump or the "Rockpile," the Santung Range, with peaks from 15,000 to 20,000 feet. After passing the Mekong River, signs of civilization could again be observed from the air. The China bases were located on the Yunanese Plateau at an altitude of 6,000 feet. [11]

The jungles of India and Burma held adversities for downed airmen in the form of insects, wild animals, leeches, headhunters, Japanese patrols, and unfamiliar food sources. The impenetrable jungle of northern Burma was dreaded by flight crews. "In that area of the world there were hundreds and hundreds of square miles of jungle so thick that crew members coming down only 150 feet apart would not be able to hear each other's cries. The thick growth simply absorbed all sounds." [12] (Tunner, p. 81)

In addition to the problems the dense jungle presented to downed airmen, there were huge mosquitos, blood-sucking leeches, tropical ticks that burrowed beneath the skin, and vicious ants. Although present, a number of potentially dangerous wild animals, such as tigers, leopards, wild buffalo, elephants, pythons and cobras were rarely seen. [13]

"On his first trip out the day he arrived in the Assam Valley, Lonnie Johnson found out first hand, about the surrounding terrain at Chabua. On the return leg from China, as the plane was on final approach, it crashed in the jungle. Johnson had been aware of a river nearby. After leaving the plane through the copilot's window, he found himself on the bank of what he thought was the river. In the pitch, black darkness of the night, he could not see that the plane had crashed in a rice paddy. Johnson had lost his gun, flashlight, and the rest of his gear.

"Even though it was August and hot, the crew was stripped, wrapped in wool blankets, and transported in a heated ambulance to the base hospital. The special treatment of heat was to prevent ticks from penetrating their bodies and invading their blood streams."[14]

If a downed airman survived the drop through the jungle canopy, he faced possible capture by Japanese patrols. At best, he might come into contact with one of the native tribes of the Burmese highlands who practiced head-hunting. Some tribes were friendly and would aid the airmen, showing them safe food sources, treating wounds, and leading them out of the jungle to safety. The Nagas, a local headhunting tribe, was fond of the British and the Americans but disliked the Japanese. Other tribes would turn the American airmen over to the Japanese for a bounty. [15]

2.2.2 China Side

More dangerous than the Hump, the China side was, for the most part, uncharted. Planes flying inside China ran the risk of crashing into the side of a mountain in bad weather. Despite the inherent dangers, air transport was the only means of supplying forward or remote Chinese positions. ATC operations within China provided the solution for the problem of quickly transporting supplies and personnel

Assam Trucking Company

across the vast areas of western China where roads were few or nonexistent.

Landing strips on the China side were unique. The runway at Kunming was made of gravel "chipped to size by hand with hammers, then carried to the site in small baskets, and laboriously spread by hand. Thousands of Chinese laborers, working with only the crudest equipment leveled the ground, covered it with hand-crushed stone and then rolled it with hand-pulled rollers—200 men to the roller." It was crude, rough, bumpy and hard on tires. The largest field required 250,000 Chinese laborers for construction. [16]

While offloading cargo in China, another fuel delivery method employed was to drain the transports' fuel tanks to the absolute minimum to make the return flight to India. The fuel drained was used by the 14th AAF (Flying Tigers). [17] A lesson learned early in a pilot's time on the Hump was that his best friend would be a stick with which to check the fuel level in the plane's fuel tanks. Sometimes it meant the difference between making it back to an Assam base or becoming another navigation point on the "Aluminum Trail." Many pilots learned quickly to closely supervise the draining of their fuel tanks, gauging their remaining fuel level on their faithful stick. There were a few exceptions.

"One C-47 crew had the opposite experience. The pilot went into operations in Kunming and filled out the required forms specifying that his plane should be filled with 2,500 pounds of cargo and the tanks half filled with fuel. The crew then proceeded to the mess shack. After eating, the pilot found his flight plan had been approved and the plane was ready for the return flight to India. With preflight and clearance to take off done, the pilot began his roll down the long dirt runway. Operations noticed that the plane did not take off at the usual place in its roll. A few crewmen standing outside Operations watched the transport and noticed the same thing. Instead the plane took all two miles of strip to get off the ground and, even then, the pilot didn't really takeoff. He raised the landing gear and roared along a few feet above the flat countryside south of the airfield. To those watching, it seemed the pilot would never make a turn or gain any altitude.

"After some minutes, when the C-47 had finally climbed to about 20 feet the pilot banked the cargo plane very slightly and turned a few degrees south. This exercise continued until the plane crossed the bank of the big lake outside the city of Kunming. The pilot expertly continued to increase airspeed and altitude until he could get back into the traffic pattern at Kunming and about forty minutes after take-off, he was able to land the transport.

"On the ground, he reported that he had run the engines at full power for minutes rather than the seconds recommended. The whole plane had been under tremendous strain. While in the air, the pilot had guessed what had happened and this was proven as soon as he walked from the plane, his uniform drenched. Two Chinese loading crews had each put 2,500 pounds of cargo in the plane. Two groups of Chinese had each had filled the fuel tanks. Thus, the plane was well loaded and fueled for a take-off at Kunming, where, because of the altitude, take offs aren't normal by stateside standards." [18]

2.2.3 Sabotage

Having made the flight to China, the aircrews faced other dangers on the ground. An armed guard was always left with a plane during unloading of cargo by Chinese laborers, who were known to place explosives on the planes in order to collect a bounty from the Japanese. [19] Marauding Japanese fighters, an ever-present danger, caused more than one Allied crew to spend time in slit trenches or, as one reported, a Chinese graveyard at night, with only a side-arm. One regular called "Bed Check Charlie" could be counted on to make an appearance just before dark. [20]

2.2.4 Chinese Chicken

The laborers, who believed a "Dragon" was following them and causing them bad luck, would run across the runway in front of planes rolling for takeoff. This Oriental form of "chicken" was played in order to kill the Dragon, thus changing their luck. Many flights were aborted, with tragic results. It was not uncommon for a transport to flip over on its back for Chinese on the runway, killing the crew and the Chinese laborer. The order was issued not to abort on takeoff for any Chinese on the runway. Some of the Chinese actually made it across. Those Chinese who did not successfully make it across the runway often caused damage to the transport, making the plane's return to India questionable if the damage was extensive. [21]

Regardless of the hazards, the airlift was the most efficient way to supply Chiang and his forces. Brigadier General Stuart C. Godfrey, chief air engineer, noted that "in bad terrain, one plane can do the work of a dozen two-and-a-half-ton trucks; planes last longer; manpower support is less with planes; and less money and manpower is needed to build and maintain airfields than roads, especially through the jungles and mountains." [22]

In the best of weather conditions, the flight east took about 5 hours. After a round trip the crews were often stiff, sore, and exhausted from the flight and its accompanying tension.

In the worst weather conditions, the toll on crews was more pronounced.

2.3 BURMA

2.3.1   Hill Tribes

Major Robert Wright, Division's Assistant Chief of Staff, Intelligence, established relations with a primitive network which had already been developed with the natives of India and Burma. A patch was designed which each flyer wore on his flight jacket identifying him as an American flyer in Chinese, Burmese, South Shan, West Shan, and Sgaw Karen. A reward was promised for the return of the crewmember. Wright kept a safe full of gold coins in his office for payment of the rewards.

Little trouble was expected among the hill tribes of northeast Assam, and no firearms were needed for self-protection. The boundaries between the tribes were fairly well demarcated, and the tribes were easily distinguished by their appearance and clothing or lack thereof.

Abors: An extremely well-built, stocky people, the Abors were very loyal to those they knew and respected. They usually wore a woven cane hat.

Mishmi: Their clothing was similar to the Abors, but as a group they appeared much dirtier. Known for their physical endurance, the Mishmis could cover long distances at altitude with little physical affect.

Rongpong Naga (Singphos or Kachin-type): They were effete, semi-civilized, harmless and easily distinguished by the Burmese "lungi"—a skirt-like garment of some sort of plaid cotton cloth—that they wore.

Naga (Proper): The Naga were a cheerful people, capable of doing hard work. They preserved their tribal customs and dances. [23]

Kohimas: Known head hunters who generally killed anything in a uniform, but tended to kill Japanese more than Allied personnel.

2.4 WEATHER AND WEATHER REPORTING

Although a number of problems had prevented the increase of airlift tonnage, weather conditions, in combination with the hostile terrain, proved to be the major rival faced by the flight crews. One crew member stated that the "weather was their worst enemy." [24]

Between the Assam Valley bases in India and Kunming, China, the air route over the Hump "passed through the turbulent meeting place of three major Eurasian air masses. Low pressure masses from the west moved along the main ranges of the Himalayas to the Hump, where highs from the Sea of Bengal collided with Siberian lows." [25] The cold, dry artic air being funneled through the mountains met the moisture-laden tropical air to produce some of the most turbulent and unpredictable weather in the world. The weather conditions changed from minute to minute.

2.4.1 India Side

On the India side of the Hump there were four seasons:
- Spring—February, March and April
- Summer—May through September
- Fall—October and November
- Winter—December and January [26]

The "little monsoons" occurred in April and were marked with a few weeks of torrential rains. The true monsoon season began in May and continued to mid-October. The usual rainfall was 200 inches per year. From the end of October until January or February the weather was better but violent storms with turbulence in February and March were frequent, with winds of 75 to 100 mph.

When the summer monsoon winds blew out of the southwest, they gathered immense quantities of ocean water, in the world's greatest hydraulic action, then swept inland over the subcontinent of India and southern Asia. The pure evaporated ocean water descended when the monsoon winds moved over land, and the most violent descent occurred in Upper Assam. The winds blew across land at close to sea level and were deflected by the steep wall of the Himalayas. At night the air masses caused temperatures to change from 140°F during the day to 55° below zero in the cockpit. The clash of air masses resulted in heavy rains.[27]

"In Assam, it got so hot that the sun beating down on the wing fuel tanks heated the volatile gasoline close to the vaporizing point. Fuel vaporization could happen, especially with the C-46 after takeoff for China, when the plane climbed so rapidly that atmospheric pressure decreased faster than the gasoline cooled off. Vapor lock often resulted in midair. When this happened to both engines at the same time, the plane was doomed, and its crew might be doomed as well." [28]

When the wind was out of the North blowing to the Southeast, a downdraft situation existed over the Hump. When blowing from the South, an updraft situation existed. A crew flying into China in clear weather would unload and be airborne again in 45 minutes. The weather over the Hump would have had a shift of just a few degrees of wind direction while they were on the ground. Because of the shift, the

Assam Trucking Company

Hump would be socked in. The variability of the weather under the conditions over the Hump made the ability to predict the weather marginal. In spite of the conditions, one crewmember felt the meteorologists did a superb job. [29]

The variety of weather in CBI ranged from the 200-inches of rain per-year, steamy Indian jungles to the mile-high, desert-like, plateau of China with its inherent dust storms. "Unfortunately, it seemed that when the weather was good in India it was terrible in China." [30]

At the altitudes required to clear the mountains, the buildup of ice on the wings and in the carburetors became a concern. Icing generally occurred above 15,000 feet except during the monsoon months of May through September, when it occurred between 17,000 and 18,000 feet. In the winter months of December and January, icing occurred as low as 9,000 feet. Without proper de-icing devices or crews familiar with proper de-icing procedures, the planes and airmen were in constant danger of ice-laden planes, loaded to the maximum limit with cargo, falling below safe minimum altitudes.

The icing, coupled with heavy thunderstorm activity, strong winds and fog, made the Hump the most dangerous airway in the world. In terms of weather, the flight from the India side of the Hump to China in the east, however, was not the most dangerous leg of the flight. The return, with strong headwinds and constant danger of downdrafts made the east-to-west run more difficult. Many planes did not complete the return flight to India. With minimum fuel and strong headwinds, it was anybody's guess if a flight would make it back to the Assam bases.

2.4.2   Hump

Due to the constant buildup of clouds over the mountains and the clash of weather systems, the air over the Hump was extremely unstable at minimum altitudes, adding to the possibility of a crash. In order to escape the turbulence caused by the mountains and weather systems, pilots had to climb to altitudes where severe icing conditions existed most of the year. The combination of turbulence, altitude, and inexperienced flight crews led to a number of crashes, earning for the Hump route another nickname—the "Aluminum Trail."

2.4.3   China Side

The China terminals had but two seasons every year—Summer and Winter. Summer occurred from May to September, and winter from October to April. In May, China's weather brought a 600 percent increase in rainfall, with a gradual increase until July followed by a gradual decrease until September. During this season, ceilings of 300 to 1,000 feet were not uncommon, with frequent closings at Kunming of one to two hours. The bad weather could last for a week at a time. Monsoon season brought more turbulent weather to the China side than to the Indian side, with only half the rainfall. Winter ushered in fog and haze with light drizzle. Icing remained the biggest problem during this season in China as it was in India and over the Hump.[31]

2.4.4   Worst Season

Springtime on the Hump was the worst season of the year in terms of turbulent winds and violent thunderstorms. The tremendous turbulence created when 100 mph winds from the southeast crashed into the side of the mountains on the Indian side created up-drafts capable of carrying a plane up at a rate of 5,000 feet per minute. Downdrafts occurred over the river valleys and on the China-side would carry a plane down at the same rate. On a particularly rough flight, Radio Operator (R/O), "Pfc Lee Ford was called upon to jettison cargo. As he was throwing out the cargo, he was tossed out of the airplane when it lurched. Hanging by a rope tied around his waist, the wind then tossed him back into the airplane." [32]

2.4.5 Worst Day

Pressured to raise tonnage figures, Colonel Tom Hardin, who was in charge of Hump operations under Brigadier General Earl Hoag from late-1943 to mid-1944, issued the dictum that "effective immediately, there will be no weather over the Hump." [33] This edict was followed until January 6, 1945, when the loss of planes and crews due to an especially severe storm caused commanding Brigadier General William Tunner to shut down the flights for ten hours. "In one day, the India-China Division lost nine aircraft, eighteen crew members and nine passengers; Chinese Civilian Transport Cooperation lost all three planes aloft that day; and, the American Tactical Command lost three planes." [34]

For the first five days of January 1945 there was good weather over the Hump. On the 5th, Chabua Traffic Control Office handled 1200 flight plans. Normally 1100 flight plans were filed per day, but on the 6th the rate dropped to 639 and 418 on the 7th. Inflight clearances, which were usually 550, dropped to 259 on the 7th of January. [35] Wing Historian Captain Joseph A. Krimm, noted in his January report that the weather on the 6th and 7th of January 1945 created the worst flying conditions ever experienced over the Himalayas.

Updrafts of 5,000 ft./min with cross winds of 90 to 100 MPH were logged. Severe icing, sleet and hail prevailed from 15,000 to 38,000 ft. altitude.

One aircraft was struck by such violent down-drafts and upset by a lightning gust that its airspeed dropped from 300 to 40 MPH and the aircraft went inverted while experiencing a 4,000-ft. rate of climb. The pilot, hanging on his safety belt, was finally able to right the plane at 21,000 ft. altitude using a method employed by primary students to right an aircraft from inverted flight. [36]

Another crew experienced a similar event. The flight was smooth one minute, then turned deadly the next when they flew from layered stratus clouds covering most of the Hump into a cumulonimbus thunderhead at high altitude. The aircraft was suddenly wrenched sideways, putting the plane in a 90° angle to level flight. For a time, they flew in the 90° bank. With the controls apparently locked in position, the crew-maintained altitude in this configuration until they hit a massive downdraft. Fighting the controls, the pilot and copilot watched as the altimeter continued to unwind. While the plane fell, an even more dire situation developed when it began to angle over even more, trying to flip all the way over on its back. Partially inverted, the aircraft continued to fly.

The pilot and copilot used all their strength on the controls, finally righting the plane. However, they didn't know how close they were to the mountain tops. With altitude their first priority, the pilot nosed the plane up at close to full power, despite the effect on their minimal fuel supply. They had fallen to 11,000 ft. The crew, by keeping their attention on staying in the air and thus alive, had managed to cross the first ridge and were over Assam, where there was plenty of space below them. If they had still been in the mountains, they would have been level with the mountain peaks. They never knew how close they had come to disaster.[37]

2.5 WEATHER REPORTING

The 10th Weather Squadron based in Delhi, located 1,000 miles west of the Assam bases, "found the reporting supplied by the meteorological department of the Indian government" unsuitable. The situation in Kunming was even worse. A single officer and six enlisted men depended primarily on data supplied through the Chinese air warning system. Because this group was more concerned with Japanese air attacks, they were less than dependable in weather observation.[38] Therefore, returning pilots were debriefed and the information was passed on to those pilots ready for another run. The situation faced by both east- and westbound pilots was at best unreliable. The weather was subject to change according to the time of the year. Sometimes the information from the debriefing did not match what the crews found when they were over the Hump. There were many surprises.

"The combination of weather and terrain would have made the Hump Airlift a difficult one even if the route had been over the middle of the United States. Actually, of course, it was located at the ends of the earth and peopled with strange and dangerous tribes, and the Japanese." [39]

2.6 ALTITUDE

The minimum altitude needed to ensure safe passage over the higher ranges of the eastern end of the Himalayas was 18,000 to 19,000 feet. Many of the transports had a service ceiling of only 23,000 feet. Mount Tali, on the China side, rose well above 22,500 feet. Most CBI pilots had a healthy respect for Tali. It is "possible that it is the only aluminum-plated peak in the world." [40]

At the high altitude necessary to clear the main spine of the Hump, flying conditions were less than favorable. Because planes were not pressurized, only the flight crew had oxygen. If available, walk-around oxygen bottles were provided for some passengers. Otherwise, passengers had to take care above 12,000 feet because of the effects of anoxia such as dizziness, headache and loss of consciousness, which became apparent within a few minutes for those not acclimated to the altitude. [41] The effects of anoxia were actually advantageous to the flight crew when transporting Chinese troops who were frightened of flying to India. Often the planes would be loaded with pack animals bound for China or the land forces of Merrill's Marauders or the Chindit forces of General Orde C. Wingate in Burma. The lack of oxygen became a plus in keeping the animals under control. Lt. Col. John M. Foster USAF (ret), stated that "it was in the best interest of the flight to gain altitude as quickly as possible so the pack animals passed out." [42]

2.7 COMMUNICATIONS

As with the other operations, faulty and insufficient communications seriously affected the efficient function of the wing. "One staff officer stated that with teletype, telephone and radio communications what there were, one got quicker results by starting a rumor. A top priority PX message to Chabua concerning the expected arrival of Madam Chiang arrived after she had come and gone." [43]

Communications over the Hump route consisted of one radio beacon at Fort Hertz, a British outpost 150 miles north

Assam Trucking Company

of Myitkynia in far northern Burma. No other guidance or communications existed. Command transmitters could not be used above 10,000 feet altitude due to icing conditions making reception weak. [44] The fields at either end of the Hump-run had voice communication capability for local flight activity only. As limited as the system was by today's standard; communications were state-of-the-art for the period. If the CBI territory had been occupied by the Allies, more communications stations could have been installed, improving the situation. [45]

Between China terminals and Indian bases communications were almost nonexistent. Kenneth G. Fransted, flying a C-46, crashed at Yunnayi when his landing gear failed to lock. The plane bellied in and, the ever-resourceful Chinese utilized the fuselage as a snack shack on the side of the runway. No one notified the crew's home base at Mohanbari of the crash or the new addition to the flight line at Yunnayi. With operations at Mohanbari ready to declare the crew MIA, Fransted and company walked in for a debriefing. [46]

Adding to the turmoil, Japanese ground patrols were known to drop 12-hour signaling devices into the jungle along the Hump-run in order to draw transports farther to the south into the range of Japanese fighters. Incoming pilots were debriefed for information concerning strength, location, and frequency of signals picked up along the route. This information, like the weather information, was passed on to those pilots being briefed before takeoff. One crew was so off course they landed in Tibet and came out on yaks. [47]

2.8 NAVIGATION

Navigation was primarily accomplished by Dead Reckoning (DR). The pilot was assigned identification "friend or foe" color codes and flares. The Loran (Long Range Navigation) [48] was available in theater, but few knew how to use it. For most it was good in the flatlands but did not work well in the mountains. One-third of the time the pilots flew on instruments.

Flights were made during daytime hours only until the middle of September 1942. Capt. John D. Payne volunteered to fly the "Rockpile" at night. With Maj. Bordene serving as navigator and using French maps, they made the flight from Chabua to Kunming in spite of the turbulence and overcast conditions. [49]

2.9 JAPANESE

The first air raids at Chabua and Mohanbari occurred on October 25, 1942. At 1400 hours 27 Japanese bombers with fighter escort bombed and strafed the airfields. At Chabua, seven planes were hit, the fuel dump was set on fire, and craters were left in the runway. A few "laborers" from Sylhet were killed. "Almost the entire group of laborers from Sylhet burned their bashas (grass huts) and left after the raid." Local natives filled the craters in the runway. A second surprise air attack, on October 27, 1942, at 1300 hours killed or wounded many of the 800 men and 500 women coolie laborers; they simply didn't have time to take cover. Material damage was light. [50]

At Mohanbari, Dibrugarh, and surrounding American airfields were attacked by 27 Japanese bombers accompanied by an unknown number of fighters. With no warning systems

Assam Trucking Company

in place, there was no notification for the 1400 attack. Nor were there any antiaircraft defense weapons. During the attack at Mohanbari, two transports were destroyed while others were damaged by shrapnel and machine gun fire. A C-47 pilot, 1/Lt. Vernon L. Scott, whose plane was being loaded at the time of the attack, elected to save his transport and cargo by taking off. After climb-out and while gaining altitude, he was attacked by three Japanese Zero's. 1/Lt. Scott skillfully maneuvered his plane to elude the fighters. In spite of his efforts, machine gun fire did hit the plane's auxiliary fuel tank and one of the 15 drums of 100 octane fuel on board the C-47. 1/Lt Scott continued with his mission and safely landed in China. [51]

On October 28th, the Japanese returned bombing and strafing as in the previous attacks. A group of enlisted men installing equipment in a radio tower noticed the laborers running, a sure sign of impending enemy attack. They left the tower, got in their truck, and headed for the bamboo jungle at the edge of the base. After going a short distance, a bomb hit about 100 yards behind them. The soldiers stopped the truck and ran to take cover. Mortally wounded by shrapnel, Pvt. Robert A. Wood was the last one off the truck. The others tried to rush to him, but were held off by Japanese strafing. When the attack ended 20 minutes later, Pvt Wood was placed in the truck and the group headed to the 95th Station Hospital. In their haste, the driver ran the truck off the road. An ambulance was dispatched and took Wood to the hospital, where he died later that night. He was buried at the American Military Cemetery at Dinjan. [52]

Slit trenches were dug at the various base units as a shelter to get away from the bombs and strafing of the Japanese attacks. If there was an attack, personnel would run for a slit trench for cover. When digging a slit trench at Mohanbari, water was hit after digging just two feet. At Lalmanir Hat (Lalman-ir-hat) the slit trenches were found to have snakes in residence. In December 1943, personnel at Chabua headed for the slit trenches, only to watch the Japanese aircraft flyover on their way to attack the CNAC base at Dinjan.[53]

As the Japanese moved closer to the Assam Valley, they were stopped at Kohima. The British were located on the west side of Kohima with the Japanese on the east side. The Kohima Tribe, Christianized by Presbyterian missionaries, helped keep the Japanese from advancing into the Assam Valley by following the patrols sent out by the Japanese commander. The Kohima Indians would strip down to a loin cloth and follow the Japanese patrol. As quietly as possible the Kohima would slip up behind a Japanese soldier and would run a finger lightly down the back of the soldier's neck. If they felt a stiff stand-up collar which indicated a Japanese uniform rather than a rolled collar typical of Allied personnel, they would cut the head off the Japanese soldier. They would then repeat the procedure with another Japanese patrol member.[54]

The Japanese had rallied and were making a concentrated push into China to capture U.S./Chinese bases that had been harassing the Japanese homeland, and into India to cut the supply line at Imphal that fed the Assam bases for the airlift effort. Japanese fighters based at Myitkyina in northern Burma harassed the unescorted, unarmed, slow-moving, and heavily laden transports. The harassment made it necessary for the transports to take the more northerly and dangerous route over the main spine of the hump.

A communication was sent by Headquarters, Eastern Sector ICWATC on 26 November 1943 to the Commanding Officer, HQ Project #8 ATC, Misamari, India, Subject: Fighter Protection for ATC Aircraft.[55] Information had been sent to all groups that maximum fighter protection was being forwarded north of the Sumpra Bum area for all ATC aircraft. That area was located in the vicinity of Ft. Hertz in Northern Burma. [56]

2.10   TONNAGE FIGURES

At the beginning of the Hump Operations, the cry had been "Ten thousand tons by Christmas." It didn't happen.

The airlift started in the summer of 1942 barely moved 85 tons of supplies to China. The CNAC planes moved just 221 tons. By July 1943 the net tonnage was 2,916; July 1944 the figures had increased to 18,975 net tons; and, by July 1945 the figures had jumped to 71,042 net tons moved across the Hump by ATC and various other units under its control. [57]

2.11   SHORTAGES

As if all the other challenges were not enough, there was always the issue of shortages. The long supply line presented its own problems as supply shipments passed through various commands on the way to India. At any given point, a shipment or part of a shipment could be commandeered if a C.O. along the way felt he had a greater need, which was usually the case and not the exception. There were shortages of trained aircrew and maintenance personnel, aircraft parts (that appeared to be caused by an emphasis on aircraft production and not spare parts), fuel, and cargo for the trip over the Hump. In the early days of the Hump Operations it was not uncommon for the rule of "first come, first served" to be in effect. There was also an antagonistic attitude between the ATC and Air Service Command (ASC), which had a disinclination to accept ATC consumption figures. Each group within a larger group was vying for supplies; each had, what amounted to tunnel vision. They simply could not see the overall mission. Supplies earmarked for ATC were sidetracked by ASC and pooled with the 10th AF supplies, including 6 DC-3s/C-47s. [58] When General Arnold intervened in January 1943, the supplies that had been purloined were recovered. [59]

There was a lack of reinforcements, as well as recreational facilities, PX and Quartermaster supplies. Adding to the lack of basics of razor blades and toothpaste was the supply of clothing, paper and office equipment. There was, however, no lack of heat, rain, poor living conditions, and malaria.

2.12 TURNOVER OF COMMANDERS

The CBI Theater was considered to be the end of the line for a commander. "The Hump was far and away at the end of the supply line. In many respects, the entire command was just plain forgotten." [60] Allied operations were very muddled until late in the war, with a tangled chain of command. There were conflicts of national policy and personnel feuds among commanders. Major General George Stratemeyer, who commanded the American air forces in CBI between 1943 and 1945, described the organization "as a strange animal with all headquarters and no hindquarters." [61] They were not there to win the war but to hold the enemy in check until victory was achieved in Europe or the Pacific. [62] It proved demoralizing for the men and their commanders.

"It was a graveyard for commanders. The ATC in CBI had in the past been considered a dumping ground for discipline problems, alcoholics, and general misfits." [63]

Assam Trucking Company

Commander	Service	Tonnage Figures
Brig. Gen. E. H. Alexander	12/1/1942-10/15/1943	28,415
Brig. Gen. Earl Hoag	10/15/1943-3/15/1944	33,076
Brig. Gen. Tom Hardin	3/15/1944-8/31/1944	85,112
Brig. Gen. William Tunner [64]	8/31/1944-9/1945	462,273

Table 2.1, ATC Hump Operations Commanders & Tonnage Figures[65]

Every man interviewed for this record said the same thing. They were there to keep the Japanese from joining with the Germans and Italians to achieve world domination. The Japanese were there to keep the British and the Americans busy so the war in Europe could go on. The American soldiers of ATC felt they were an important piece of saving the world, not the misfits and discipline problems alluded to by General Tunner. Their view was from the bottom up, while Tunner saw it from the top down.

2.13 NUMBERS LOST BY ATC

Aircraft	Crashes	1943	1944	1945	Dead
C-46	152	34	71	47	95
C-47	17	6	6	5	17
C-54	4			4	9
C-87	31	6	16	9	18
C-109	16		4	12	22
TOTAL	220				161

Table 2.2, ATC Aircraft Losses[66]

At the end of 1945, Search and Rescue noted the following: Of the 509 closed crash sites, 328 were ATC planes. There were 81 crashes still open. Within those crashes, 1314 men were dead, 1171 were able to walk out, and 345 men were still listed as missing. "The majority of those who were found dead died in the crash of their planes or in making their parachute landings. Some died later of injuries or disease, a few were caught and executed by the Japanese, while a very few fell victim to hostile natives." [67]

2.15 AWARDS

Most awards for valor and achievements were given for a particular event. Even though ATC in CBI was considered a non-combat operation, the pilots flying the Hump were in unarmed and unescorted transports over the most

Assam Trucking Company

treacherous country in the world. Later they were considered to be flying under combat conditions 100% of the time. All of their flying hours were, therefore, totaled for awards of the Air Medal (250 hours) and the Distinguished Flying Cross (500 hours). Each additional 250 hours of flying time earned the pilot an Oak leaf cluster for both medals.[69] The Air Medal and Distinguished Flying Cross were given to bomber and fighter crews flying over Europe and in the Pacific, but not to transport aircrews. The men of ATC in CBI were considered a different breed.

Chapter 3.0
- Aviation Cadet Training

In 1938, with war imminent, various Army Commands established programs to prepare large numbers of men for induction into the officer corps. Known as "90-day wonders," most were recent college graduates. After induction, they quickly completed basic training and Officer Candidate School (OCS) for US Army service. As air power rose in importance, the Army Air Corps established their own OCS training center in San Antonio, Texas. In 1940, to meet the demand for military pilots, cadet flight training was reduced to seven months, with only 200 flight hours required for graduation. [1]

To fill the gap until a sufficient number of the "90-day wonders" were ready, the Army training program targeted other resources, including commercial airline and other civilian pilots. The initial group of commercial pilots consisted of reservists recalled to active duty. In June 1941, many of the unattached civilian pilots available for training were bush fliers, small commuter airline operators, test pilots, stunt fliers, crop dusters, barnstormers and individuals with private pilot licenses. Each prospective trainee was required to have a minimum of 500 hours flying experience. Depending on the need for pilots, this requirement fluctuated to a low of 200 hours. By 1944 the requirement had risen to 1,000 hours when the US Army Air Force training program was in full swing. [2]

The first pilots in CBI for ATC were service pilots, civilians who were usually airline pilots. They had an "S" on their wings. They received pay, uniforms, and housing, but were not allowed to fly into combat areas. They were euphemistically referred to as the "Army of Terrified Civilians" (ATC). [3]

The US Army Air Corps became the US Army Air Force (USAAF) in June 1941. Cadet flight training became the Aviation Cadet Training Program (AvCad), and the grade of Aviation Cadet (AC) was created for pilot candidates. As the war progressed, several programs using various methods of training were tried and shelved or overcome by events (OBE). Training of aircrew members was more frequently tied to availability of transport planes being delivered by the manufacturers, e.g., Boeing, McDonnell Douglas, and Curtiss-Wright. "Chronic shortages of training planes as well as other equipment, coupled with the number of instructors, was inadequate." [4] Sometimes the cadet found himself "hurrying" to get through a particular phase, only to find himself waiting for a billet to open in the next phase.

To be able to apply for Aviation Cadets, a young man had to submit a written application. As with other commands, the prospective candidate needed to have completed at least two years of college or three years of technical or mechanical training. But by 1942, with manpower requirements rising, an aviation candidate with no college could submit an abbreviated application package that included:

- One application, notarized
- Birth certificate (with seal)
- Three letters of recommendation
- Parent's consent form if under 20
- Medical questionnaire
- Signed statement of no previous examination (mental or physical)

The applicant was given further instructions that his submission should be neat, all signatures must appear as stated on birth certificate (no exceptions), and all papers were to be made out in ink or typed, with typed being preferred. The applicant was to remember that "he is making application for a commission as an officer in the Army of the United States." [5]

Shortly after Pearl Harbor, applications for aviation cadet training numbered in the thousands, swamping the system. More training bases appeared across the state of Texas, under contract with private flying schools. The Army supplied the students, training aircraft, flying clothes, textbooks, and equipment. The contractor supplied instructors, training site, aircraft maintenance, quarters and mess halls. [6]

One prospective aviation cadet, just 18, applied after graduation from high school in May 1942, and was placed in a "holding pattern" with exempt status as a Clerk Typist (Civil Service) with the Army Air Corps Supply, Air Depot Detachment in Tulsa, Oklahoma, from July 1942 to January 1943. On Nov. 10, 1942, he was inducted into the Army Air Corps as an Aviation Cadet. [7] Between signing the papers and reporting for duty in San Antonio, TX, the AC and his high school sweetheart were married at Christmas. After the first of the year, he would be off to basic training. Because while in basic he would not be able to get an overnight pass to see his wife, it was decided she would return to high school to

Assam Trucking Company

finish her senior year. She joined him after her high school graduation, moving from air training camp to air training camp until his assignment overseas.

In World War II America, men were moving from training base to training base after basic training was completed. It was during this time that the men could get weekend passes to spend with their wives or girlfriends who followed their husbands or boyfriends from camp-to-camp. Like so many other wives before and after her, AC Foster's wife became a "camp follower." She had a small suitcase in which she had packed all she needed. They didn't have much money. Her allotment was $22/month, and her ration stamps were limited.

Out of the allotment, she had to pay for her lodging, food and transportation. She would find a rooming house close to the base and, if possible, a part-time job at a dime store or soda fountain. Like other wives, she wanted to become pregnant before her husband deployed—to have a child who was part of her husband in case he didn't make it home. Following him took its toll physically. Because of limited funds, she didn't always eat three meals a day—sometimes only one, if she was lucky. Soon after she suspected she was pregnant, she miscarried. She had to keep it to herself. To tell her husband that she lost their baby before he even knew she was pregnant could affect his training. The other wives proved to be a support for her as much as she was for them. When Foster finished all three phases of training and went on to the Operational Training Unit in Reno, his wife returned home to Oklahoma to wait for him and the end of the war.

The wives forged new friendships that would continue long after the war ended. They kept in touch as they moved through the training cycles. While their husbands were in class, the wives, if they didn't have a job, found their time filled with letter writing and playing bridge—*a lot* of bridge.

On January 28, 1943, AvCad John Foster arrived at the San Antonio Aviation Cadet Center to start his basic training, followed by aviation training. His entry into the United States Army Air Force (USAAF) began with two weeks of classification training to determine whether he would train as a pilot, navigator or bombardier. Foster was classified as a pilot and continued on to Pre-flight training.

There were a number of colleges across the country that established College Training Detachments (CTDs). For Foster, Pre-flight training took place at the 94th CTD at Southwest Texas Teacher's College in San Marcos, TX. This Pre-flight training was broken into two parts. At Southwest, the 2-month curriculum covered a variety of subjects from physical training and sports, academics (physics, history, geography, English, mathematics), and medical aid to the basics of flight through application of aeronautics, deflection shooting and thinking in three dimensions. A ten-hour evaluation was accomplished in a crude flight simulator (Link Trainer) called a "blue box," followed by a ride-along one-hour flight with an instructor. The next three segments of training, Primary, Basic and Advanced, typically took three months each to complete. Those who passed were awarded Cadet Wings and promoted to Primary Pilot Training.

Upon graduation, Foster became a member of Class 44-F in Primary Pilot training at Bruce Field in Ballinger, TX. Primary Pilot training included 60 to 65 hours in a 2-seater aircraft with an instructor. At any one level, if a cadet "washed out," he was sent to navigator or bombardier training. On December 31, 1943, J. M. Foster, soloed.

The objectives of Primary Training in accordance with The Student Pilot Handbook, Hunt and Fahringer, 1943, were to:

- "Give the student a solid foundation for becoming a competent military or commercial airline pilot. To turn out a man who

    o   Uses his initiative under flight conditions and not be satisfied with just doing a passable job
    o   Shoots for a perfect score in whatever direction he aims.
- Help mould the qualities of an officer and leader into each student; to train men capable and willing to take orders properly given and carry them out to the letter without excuses or alibis." [8]

The students learned to work with instructors as a team. The Student Handbook became a pocket-size quick reference the student would carry throughout his military service.

After completing Primary Flight Training, Foster transferred to Basic Pilot Training at Goodfellow Field in San Angelo, TX, where he learned aerial navigation or instrument flying, and night, distance, and formation flying. (Over the four years Goodfellow was in active service, more than 10,000 pilots were trained at the Army Air Corps Basic Flying School.) Foster's last move before graduating from Cadets was to Pampa Army Air Field in the Panhandle of Texas. Activated in August 1942, Pampa Army Air Field served as the location for advanced twin-engine training. On August 3, 1944, Foster graduated from Aviation Cadets as a member of Class 44-G and was commissioned a Flight Officer. The Flight Officer rank was a Chief Warrant Officer, Junior Grade (CWO) rating established by the government with Public Law 658, Flight Officer Act, passed on July 8, 1942, for those new pilots with no college, not at the top of their class, or for enlisted applicants to the AvCad program. The new pilot was neither enlisted nor an officer. He was in the middle of his class, and could gain rank to 2nd Lt. with time and experience.

Even though he was a Flight Officer (FO), after completing pilot training the AC was rated as a pilot. [9]

3.2 Operational Training

Generally, pilots with a 2-engine rating could go to bomber training or transport for replacement of pilots overseas. The replacement requirements for heavy bomber aircrews in both the southwest Pacific and European theaters reduced the number of available pilots for the Air Transport Command (ATC). The number of aircrews needed was tied to the number of heavy bombers produced per month. That meant 1 pilot, 1 copilot, 1 navigator, 1 radio operator, and 1 flight engineer per crew. An estimated 16 bombers produced per month required 48 officers and 32 enlisted men minimum, but actually twice that number of aircrew members were trained. [10]

From Pampa there were two more stops Foster would make before deployment. Transition flight training for multi-engines began for him at Majors Field in Greenville, TX, where he found himself at one point pressed into service as an instructor. After a month at Majors Field, Foster went to the 585th AAFBU (Army Air Force Base Unit), 3rd Operational Training Unit (OTU) in Reno, Nevada. The OTU was designed to give an individual training to become part of a team. Foster flew as one of a three-crewmember team to gain experience on the C-46 at Reno.

The 3rd OTU was the final step in preparation for flying the Curtiss C-46 Commando aircraft for the Air Transport Command (ATC) in China-Burma-India (CBI). The mountainous terrain between Reno, NV, and Fairfield-Suisun Army Air Base (AAB), CA, (now Travis AFB just to the east of San Francisco) was similar to the Hump terrain. Often, the crew would train with cargo onboard similar to the type and weight the aircrew would encounter, allowing them to develop critical handling skills they would need when flying the Hump. [11]

Two other OTUs were in operation to prepare aircrews on multi-engine aircraft. The 1st OTU, established at St. Joseph, MO., trained aircrew members on the C-47 transport; the 2nd OTU at Homestead, FL, trained aircrew members in the C-54, C-87, and B-24 specifics as a crew. Flying conditions were fashioned to reflect the flying conditions the aircrews would face in their assigned theaters.[12] All these airframes were employed over the Hump.

Upon completion of the OTU, Foster was sent to Nashville and held in a secure area (secret-level orders) until he could be shipped out to his wartime assignment. From Nashville, he went to Miami as his departure point for India.

3.3 In-Theater Training

Upon deployment Foster, like all aircrew members, found his training would be an on-going exercise. Once qualified as a pilot, copilot or radio operator, the crewmember was expected to maintain flight qualification in their crew position through a series of exercises known as currency training, including review of checklists, technical manuals, and operational procedures of concern for the various aircraft in their particular theater and assigned base unit. In some cases, time would be spent in one of three Link trainers (simulators) available in theater. The Link trainers were installed in basha-type buildings similar to the buildings used for personnel housing. A Link Trainer arrived at Misamari 1326th AAFBU on March 1, 1945.

For the trainers, transition curriculum was developed to include a cross-country flight to China via the "Easy" route with the return leg via the "Charlie" route. The "Easy" route was an eastbound route that took the pilot from Chabua AAFBU in the Assam Valley to Kunming, China by way of Myitkyina, in northern Burma, a more southerly route available after the Japanese were driven south in Burma. The Charlie route was the westbound leg from Kunming to Chabua, flying north of Myitkyina on a more direct route.

Flying the practical exercises in the trainer aided in an increase of pilot proficiency. An advanced curriculum was developed with a:
- Cross-country flight to China via "Fox" route
- Localizer landings
- Simulated let-down at Chabua with use of localizer

The pilot would "fly the beam" on approach and would listen to a series of dits and dashes to gain direction of the glide path. • – (A --to the left of the glide path) and – • (N – to the right of the glide path). A continuous noise (no dits or dashes) indicated the pilot was "on the beam."

Procedures of concern included takeoff restrictions to 300 feet in good weather (go-around? or landing.)

If the engines quit at an altitude of 500 feet, the chances of getting out alive were slim. "Our operation procedures should be predicated on the fact that if there is a mechanical failure between that and 700 feet, there is not much that any ruling could do, but the airplane should be able to climb enough for the crew to 'bail out' if necessary." General Alexander set the takeoff ceiling restriction at 500 feet "until certain maintenance is good enough to make it less." [13]

In spite of the trainers and better charts, many of the pilots used crash sites as points of navigation. Any new crash sites

Assam Trucking Company

were reported on landing during debrief. This form of navigation earned the Hump Route the infamous nickname, the "Aluminum Trail."

Pilots completed ground school and proficiency exams as part of their currency training to maintain skills and current engine/aircraft ratings. Copilots completed transition training for upgrade to 1st pilot or reserve 1st pilot status.

For Radio Operators (R/O's), the salvaged fuselage of a C-46 was moved to a new location at Misamari to be used as a radio shack. With radio equipment installed, the salvaged wreck became a well-equipped schoolroom for initial training of new R/Os (always in short supply in-theater) and refresher/continuation training for experienced R/O's. The new LORAN (Long Range Navigation) system, operated by navigators in other theaters, was used by the radio operators in CBI. [13]

At Misamari the education program included crew currency and qualification, indoctrination of new personnel to the theater, flight conditions and peculiarities. A booklet, ATC in India-China, produced by the India-China Division was given to new personnel. Restricted until September 1945, the booklet reviewed the history of the Hump flights, how and why they were started, plus short profiles of some of the command personnel. Some local customs were covered, e.g., that cows were considered sacred by Hindus, laundry practices, and that there are two India's (British India and the Indian States). Even an ATC weekly paper, The Dragon, in addition to providing local news and entertainment, was used to further educate military personnel on information dealing with indigenous snakes, educational opportunities and safety issues. Training was of vital importance in CBI.

3.4 Jungle Indoctrination

The DECLASSIFIED syllabus, "Vacationland in the Jungles of the Himalayas," outlined what the crewmembers would likely experience. To alleviate fear of the jungle and to integrate relaxation with training, Jungle Indoctrination camps were established for each base unit. This training was considered a must for aircrews, especially if they survived a bailout and could walkout. The vision of the Jungle Indoctrination camp at Pathalpiam, in particular, was that of Col. Harry Renshaw of Nogales, Arizona. In November 1943, Col. Renshaw activated the camp as a place where military formalities were suspended. Rank was not recognized during the officers' and enlisted personnel's stay. Officers could stay 4 days, while enlisted personnel stayed 7 days. At the camp on the banks of the Kemi-Nodi River (Lazy River), the entrance to the dayroom was decorated with game skulls and other prizes of previous visitors—a reminder of the reason they were there.

Equipment required to conduct training included:
1. Salvaged parachutes
2. Emergency parachute kits
3. Jungle knives
4. Compasses
5. Bamboo (large)
6. Signal mirrors
7. Literature
8. Map of the area
9. Transportation
10. Insect repellent
11. Bulletin board

Even though the camp was seen as a break from the monotonous schedule of sleep-fly-eat of the Hump flights, the work the aircrews did there was very serious by nature. The skills to be mastered just might save their lives'. The syllabus

given to those who came for the training outlined in detail how each day would be spent, emphasizing how serious the training was. Participants on Day 1 of the syllabus were taken by a guide into the jungle and taught how to:
- make and use utensils from bamboo
- identify plants and animals to both eat and avoid
- track animals
- set up a jungle camp
- cut through dense jungle growth
- use alternate sources of water

A river trip focused the trainees on available food sources, how to read the river, and how to tie a raft above the high-water line. They were taught how to approach a native village, how to conduct themselves while in the village, what to pay for food or other articles, and what gifts to give.

The men were told they had three days to complete the training, which ended in a jungle trip without a guide on the third day. Intense but necessary, at the end of the training the participants were expected to survive after bailing out into the dense jungles that existed along their flight route over the Hump between India and China.[14]

Besides training, a rest and relaxation camp was located at Shillong, India. Because the Assam Wing had been requested to supply trained men urgently required in other areas, those retained were given additional duties to release as many as possible for reassignment. A large group were sent to the AAF Rehabilitation Center in Shillong. There was a quota for flight and ground personnel, so the slots included were distributed proportionately according to strength and needs of various bases. Activities highlighted were hunting, fishing, swimming, eating and sleeping.[15]

2/Lt Foster reported that, as part of his R & R, his group went on a tiger hunt. They built a platform in the trees, staked out a goat, hung a lantern, climbed up to take their place on the platform with 30/30 rifles and waited. The tiger secured the goat before they could react. They tried a second and third goat with the same results. Foster said they lost three goats and never saw the tiger! [16]

3.5 Maintenance Training

Few trained and experienced maintenance personnel were sent to CBI, making training a must. The decision was made to hire locals to be trained as maintenance workers. Those chosen needed to speak and understand English and had to have some education. Most available were unskilled or semi-skilled workers. Despite the limitations, 986 civilian workers were hired and trained at Lalmanir Hat, AAFBU 1326th by 1945. [17]

In March 1945, ICD letter 353/230 from headquarters ICD-ATC directed the use of more civilian employees where possible. Training programs were established for locals in apprentice airplane mechanics. At Tezpur, Maj. Robert C. Thomas of Base Engineering was selected to supervise training for those cleared. Forty-five locals were cleared for classes, with five in each class and at least one able to speak and understand English. At the same time a similar class was begun for Black personnel on the field to train them for "upper echelon or more complicated maintenance with the emphasis being placed in freeing Black [18] personnel loading planes, general duties, etc., utilizing them to the greatest extent possible within the limits of vacancies and qualifications. Of the 208 Black troops originally assigned, one-third have been assigned to more responsible jobs. Originally used to load and unload, planes they were being treated as if they were nothing better than a coolie." [19]

Assam Trucking Company

At Chabua the reaction to the March ICD letter 353-320 led to the decision that, because of a 25% absenteeism rate among the locals and difficulty in understanding English and following orders, Indians would not be used as mechanics. They should instead be used for general duties allowing "white and colored" personnel to do more complicated tasks. "Coloreds should be trained as mechanics as they have shown much interest and their work is excellent under supervision in Production Line Maintenance (PLM) and engine changes." [20]

Because of the special problems associated with maintaining the C-46, it was critical that the systems be understood. U.S. Army personnel became the focus for training on the C-46 systems, because the C-46 was there to stay. Available training aides were used by some units, while others made their own full-scale mock-ups of the C-46 engines and systems.

Chapter 4.0 — AAFBUs

With the supply route to China decided, determining the location for bases of operation and the start of the construction of airfields rose to the top of the list of considerations. The Japanese push into Burma required a more northerly route. China terminals were not a problem, as considerable aerodromes already existed. The area chosen on the India side, the Assam Valley, which was known for its tea plantations, had no infrastructure to support military capabilities. The jungle and the rice-paddy world of India had been chosen to serve as the base of operations for a command that had not existed before in WWII. "Command Chaos" reigned supreme. Making sense of priorities, difficult at best, included where to build, how to address supply logistics, and who was in charge of what.

Initially, responsibility added to the trouble with the development of the aerial supply service. Like other aspects of problems in CBI, the construction of the fields was planned by the Americans and built by Indian labour under British direction. The creation of five forward and three rearward fields, to be ready by May 1942, with three more fields to be

completed by October, fell the way of other problems in the command.[1] By May, only two of the original eight fields were operational and available to ATC.[2] By agreement between the U.S and British commands, six bases with concrete strips would be built on British-owned tea plantations. However, only four bases were built by British engineers.
- Chabua—in operation 1942—Hattialli Tea Estate
- Mohanbari—in operation 1943—Greenwood Plantation
- Misamari—in operation late 1943
- Sookerating—in operation 1944

4.1 CONSTRUCTION

Tea plantations in the Upper Assam Valley, which had been chosen to serve as the foundation for the airfields, had existing supply lines—albeit by multi-gauge railroads and barges. Owned by the Crown, tea estate land-usage presented no obstacles. With the sites selected, Mr. Chapman, executive engineer of the Burma service, conducted surveys to determine the locations for the airfields on each of the plantations. Mr. Franklin, Mr. Pierce, and Mr. Paget, tea estate managers, were involved in the planning and construction of the fields adjacent to their respective tea estates. Once Mr. Chapman laid out the boundaries for the runways, plans were formulated and construction was begun in February 1942.

Construction would not be an easy task at any of the chosen sites. With no heavy building equipment available, all work had to be done by local laborers, who proved to be slow and unresponsive to the British overseers. Only six trucks were available to haul construction materials, including stone for the runways. All sites, originally heavy jungle with strips of swamp, had to be cleared, and rice paddies (underwater) filled and brought up to elevation. To add to the incongruous situation, rainfall, normally between 70 and 100 inches per year, was excessive for Assam in 1942; native laborers, panicked by Japanese bombings, fled the region in throngs, and construction equipment failed to arrive on time.

Working under primitive conditions to clear the runway site at Chabua, local laborer consisting of 3,000 men cut away the undergrowth, while 2,600 women broke stone for the runway surfaces using wood hammers while squatting on their haunches under the protective shade of bamboo sheds. "It is estimated that the women crushed 11,000,000 feet^3. Minor flooding, caused by the frequent rainfall, caused the stone-crushing site facilities to be moved several times. A mechanical stone crusher, received in April, broke down immediately with no replacement parts available—an ongoing problem in theater. The more reliable human labor continued with an additional 2,000 laborers imported from Sylhet, 300 miles southwest of Chabua." [3]

The ground in front of the Chabua Operations building was so swampy that it seemed to be a bottomless pit because of the way it appeared to swallow rock. When ready, the single runway, just 6,000 ft. long, was tarred by hand using laborers. Work finally stepped up when an American truck with a tar sprayer arrived. The first C-47 landed at Chabua on July 28, 1942. The first control tower at Chabua was perched in a tree and operated by ATC and RAF (Royal Air Force) personnel.

With the construction of dirt roads in conjunction with the runway, buildings emerged. Mr. Chapman said upon completion of the runway, "Every field I've built has been captured by the enemy. I hope the string is broken." His wish was granted; Chabua was never captured, despite the fact that it was located just 0.5-hour flight from the advanced Japanese bases.

Putting all the pieces together did not happen smoothly or overnight. As construction continued, air traffic was handled at the CNAC base at Dinjan. Messages and weather reports were sent from Dinjan to Chabua by the Signal Corps. Field telephone and administrative messages were rushed over the rough, unpaved Assam trunk highway by jeep. The communication unit was moved to Chabua in mid-September 1942, the same month the Americans installed anti-aircraft Bren guns at the base.[4] In November 1942, the following message from the General Supervisor of the Dibrugarh Sub-Division was received:

"TO:

G. E.

Chabua Aerodrome

My dear Ford,

Paget has reported the two following matters which should definitely be taken up with the authorities concerned.
1. The American AA Gunners had a firing practice on Friday evening at 5 P.M. while the labour was still working and NO warning was given that a practice was to be held. They must learn to co-operate if work is to proceed smoothly.
2. At least two steam rollers, if not more, have been severely damaged through the ignorance of some of the Pioneers trying to start them up. As you know, we have two highly qualified engineers from our organization helping you to look after your mechanical plant, and the running of the machinery should be left entirely in their hands.

If you would prefer not to have the use of these gentlemen, I am sure Paget will be able to make other arrangements. Salaams"[5]

The laborers working on the fields knew that the anti-aircraft guns were used during air raids. If used during air raid drills without giving the laborers warning, the laborers panicked and ran for cover, causing a great deal of turmoil. Some would not return.

Unfortunately, due to shortages of building materials and maintenance supplies, the British were unable to meet their contracted building commitments. Despite construction delays and shortages, the tea plantations remained in full operation during the war effort.

4.2 AAFBU ACTIVATION

4.2.1   Mohanbari, 1332nd AAFBU

Even though the construction of the physical bases in India was not completed, activation of the base units started. The 1332 AAFBU (Army Air Force Base unit) Mohanbari, officially activated at Pope Field, Ft. Bragg, North Carolina, on March 7, 1942, was part of the 12th Transport Squadron, 60th Transport Group from Westover Field, Massachusetts. This group of two officers and 79 enlisted men became the nucleus of the 6th Ferrying Squadron of the First Ferrying Group. Their mission was to carry supplies for American and Allied troops to the battle lines of China, Burma, and India (CBI).

The advance party for Mohanbari left Karachi on the 6th and 7th of October 1942. The remainder of the base unit personnel followed on the 27th of October, traveling from Karachi by train. They faced a trip of over 1,000 miles on varying-rail-gauge trains. Great care was taken as the train route passed through Hur territory. The Hurs, a band of tribesmen, spread their own brand of terror through the Sind desert. In addition, other known bands of outlaws similar to the Hurs created problems. In the central Indian provinces, the independence movement Congress party rioters disrupted travel, requiring even more care. Travel was slow, with frequent re-routings and extra precaution being taken by U.S. Army and railway officials. At Pandu, the train was stopped while Indian officials made sure it was safe to travel at night. [6]

Just a few hours into the trip, the group learned that the Japanese, in the middle of the day on 25 October 1942, had bombed and strafed the town of Dibrugarh and adjacent American airfields.

The remaining personnel arrived at Mohanbari from Karachi on 5 November 1942. With no administrative personnel available at Mohanbari, many problems quickly arose including.
- Low morale
- Shortage of planes
- Difficulty keeping up with what planes were there
- Overage of maintenance personnel

The recently arrived personnel were put to work on base construction and the usual squadron duties, e.g., policing the base unit, KP duty, etc. This was not the work for which they were trained, thus driving morale farther down.

Mohanbari, like Chabua, was carved out of a tea plantation. Done in just two months, it had enough taxiways and runways for limited operations. Tea planter bungalows housed officers. Buildings for mess halls, barracks, and work rooms were woven bamboo siding with thatch roofs. Some conveniences were added as soon as possible, e.g., plumbing in wash and shower rooms, hot water heater, and portable generators to supply limited lighting.

The Operations tower was built but not being used because of the October air raids by the Japanese. The "tower," consequently, consisted of two wooden packing cases, one on top of the other. The operator sat on top of the boxes with an umbrella to shield him from the sun.

4.2.2   Chabua, 1333rd AAFBU

While working out the kinks in the airlift effort, Chabua became the main point-of-entry and the embarkation terminal for the duration of the war. By early 1944, the Transient Center at Chabua was located on the polo grounds and included billeting, mess hall, and a large modern PX. Indian Bearers wore red turbans— "The Only Red Cap Service in India." Over 6,000 transients were fed and housed each month, from privates to generals, USO performers, missionaries and war correspondents. Nine thousand Chinese troops trained in India came through Chabua. [7]

4.2.3   Misamari, 1328th AAFBU

The 1328th AAFBU, organized at Seymour Johnson Field, North Carolina, in September 1943 as Project 8, was assigned to move pipe for construction of a gasoline pipeline, originally to parallel the Ledo Road from Ledo, India, to China.

Identified as the commander for Project 8, Major Robert M. Wilson arrived, with the first complement of 4 officers and 47 enlisted personnel assigned, to find buildings in place and the 24th Airways Detachment there to greet and in-process them. Accommodations included rope beds or charpoys with mosquito netting, pucca bashas with thatch roofs, cement floors, and screened windows. [8]

The rest of the military personnel assigned to Project 8 sailed from Charleston, SC, on the 29th of September 1943 on the Louis Pasteur (a former French passenger liner). After 8 days at sea they docked at Casablanca, then flew on to Misamari. The project's full complement of 334 officers and

Assam Trucking Company

959 enlisted personnel had 16 C-47s and 40 C-46s assigned. Two airfields in India, two in the Kunming, China, area and one at Ft Hertz, Burma, were identified as part of Project 8. Twelve C-47s and 30 complete crews were given the mission to transport steel pipe lashed under the fuselage of the transports from a designated pickup point to the extreme eastern end of the Assam Valley and to Ft. Hertz. By the time the Project aircraft, personnel and equipment arrived in Casablanca, the Japanese had invaded Thailand and had moved considerable weaponry to India. [9]

The plans were deemed impractical, and those assigned to Project 8 became part of the India-China Wing (ICW) of the ATC stationed at Misamari. While on the ship, Maj. Grevemberg, commanding officer, and Maj. Reilly, executive officer, who led the group, arrived at Misamari to find they were not in charge at all, but Majors Wilson and Townsend were. Majors Grevemberg and Reilly were the first two transferred to other bases. [10]

A second group assigned to Misamari included 782 enlisted personnel of various MOSs (Military Occupational Specialties/Work codes) who reported to Goldsboro, NC, for transfer to Misamari. They came from five other commands, or areas including:
- 379 from the Ferrying Division
- 209 from the Caribbean Wing
- 51 from the West Coast
- 137 from the Domestic Transportation Division
- 6 from Camp Luna, Las Vegas, NV

There were some sergeants among this group, but most of the men were lower ranking enlisted—privates, privates first class and corporals. The unit lacked experienced personnel, which greatly inhibited the advancement of base operations. The isolation from reliable supply sources, resulting in lack of food and equipment, negatively impacted progress at Misamari.

Because they were not part of Project 8, they did not receive the same welcome. There were no qualified staff members to orient the new personnel to their surroundings and mission, the base was in a state of confusion normally associated with the startup of any organization, and there was a lack of clearly defined command roles.

Whoever was responsible for orienting the runway at Misamari had done a stellar job of creating even more chaos because the:

- Checkpoint, at the northeast end of the runway, was just 150 feet off the Gabharu River.
- Tower was not in operation, requiring the approach to be made by dead reckoning
- Three trees at the edge of the runway were barely missed by planes on takeoff;
- Railway ran parallel to the runway, causing planes on landing to narrowly miss the invariably overloaded freight cars.

The Crash Ward at Misamari was placed on the flight line and was manned 24/7 with 3 men on hand in 8-hour shifts. Two men sat in an ambulance, while one remained in the ward. The ambulance posted at the ward was not to be moved unless there was a plane crash. In December 1943 a jeep was added, which was to be used for the dispensary in lieu of the ambulance. The dispensary capacity was 23, with an average of eight patients per day. Due to the lack of appropriate lighting, all medical operations were performed at Jorhat or Chabua and not at Misamari.

Once the two-story control tower was built, it was used to control taxiing, takeoffs and landings; however, it did not provide a clear view of the field. An additional floor was requested, but the British Garrison and the American Area Engineer refused to grant the A-1 priority needed to authorize construction. As a result, the additional floor was not completed by the end of the first week of January 1944.

4.2.4 Sookerating, AAFBU 1337 & Tezpur, AAFBU 1327

Tezpur, originally organized in 1942 and constructed by the British Royal Indian Air Force as a British base, later housed C-46 Curtiss Commandos and B-24 Liberators.

Opened in the Assam Valley in late 1944 as a sub-field of Chabua, Sookerating was used as a transport base by ATC and as a 10th AF combat airfield. On January 14 and 15, 1945, Deragon Field was closed for Hump traffic, and all equipment and personnel were moved to Sookerating 48 hours ahead of schedule. Because of their efficiency, General Tunner congratulated them for their effort and requested full details of the move, which were used to implement a Standard Operating Procedure (SOP). On 27 Jan. 1945 the 3rd ATS was attached to the 1337 AFFBU at Sookerating.[11]

4.3 LIVING QUARTERS

In general, the first crews coming into the theater found living quarters sparse. In addition to tea planter's bungalows for senior officers, others, along with enlisted personnel and junior officers, were housed in barracks constructed of pucca walls, thatched roofs and cement floors, or in British double-walled tents. The British tents were more suitable for the tropical climate than the American tents, which were referred to as "Turkish Baths." The four-man British tents were double-walled and double-topped, with an air pocket between the layers, which kept the heat from penetrating the tent. The tents, received at the end of 1944, were placed on concrete pads to keep insects under control. The better British tents were still ill-ventilated, hot and noisy, and served as housing for two-thirds of personnel. [12] Those scheduled to fly at night were unable to sleep in the canvas tents during the day due to unbearable temperatures and humidity. Upon arrival, all personnel started in the tents and were added to a list for assignment to a basha, the preferred living accommodation.

Gordon Leonard recalled a fight between two cats, native to the tea plantations, caught between the double-walls of the tent. The noise brought visions of two immense tigers fighting between the two walls. To protect himself from the bad-tempered felines, Leonard tucked his mosquito bar in more tightly around him, giving him comfort that he was better protected from attack by a tiger. [13]

Bashas, made of woven bamboo walls and grass roofs, had a dirt floor, a front porch with a rail, five windows (two in front, one in back and one on each side), which were screened and covered with a flap. At some bases the floors were brick, raised above the surface of the surrounding ground. At Jorhat and Kurmitola, the bashas were closest to the homemade showers and mess hall and furthest from the latrines, which tended to smell badly when the rains came and the latrines overflowed. [14]

The bashas, which proved much cooler than the tents, housed four men. Each man was assigned a servant/bearer to keep the beds made, the tent or basha clean, clothing mended, shoes shined, and the bedding aired.

Each bed had a mahogany frame with woven bamboo springs that stretched in wet weather. A cotton mattress, 2-3 inches thick and preferable to the mattresses made of rope by the locals, could be purchased on the local economy. After several weeks, they tended to disintegrate in the middle and became nest-like. When a crew was reported missing or down, everybody scrambled to the missing crewmembers' bashas and took their mattresses. The number of mattresses airing on the porch, proved to be a matter of honor or seniority. [15]

Indigenous snakes including Burmese pythons, kraits, and cobras, could be found in bedding, shoes, or in the rafters of the bashas. One Humpster stated that the Burmese python tended to travel in family units. The natural hazards, noted on base at Misamari, when walking across fields (rice paddies), were leeches and various snakes. It was better to walk on the road. Even though it might take twice as long, it was found to be infinitely safer.

Construction was never truly finished before the end of the war.

4.4 HEALTH & SANITATION

Despite the planning and seemingly similar building conditions, the various bases were very different, yet in some ways very much alike. The environmental issues were basically the same. Built on the edge of bamboo growth, jungle, and rice paddies, the ground was, for the most part, underwater, if only a few inches. The standing water provided an excellent breeding area for mosquitos. Because of the environment, diseases such as malaria and dysentery, were common to all bases.

Implemented to address health problems, several policies were impossible to enforce. Indian natives, who were considered uneducable, would bring flea-ridden cattle on base, and cattle, being cattle, would defecate in or near water supplies or near mess halls. Due to rice paddies, malaria became epidemic and posed a real threat. Diseases prevalent in CBI besides malaria and dysentery included small pox, chicken pox, meningitis, mumps, infectious hepatitis, and venereal disease. [17]

Malaria was such a problem at Misamari in June of 1944 the following preventive steps were initiated:
- Moved native carriers off base
- Put malaria disciplines into effect
- Increased DDT and oil spraying of mosquito breeding areas by airplane

The results noted that malaria cases dropped to a record low of just seven cases in January 1945.

New personnel were given a directive for personal discipline:
- Discipline starts at 1800 hours
- Your clothing at this time, and until dawn, will feature the following:
    - Long trousers
    - Long sleeves
    - Shirt fully buttoned
    - High shoes or boots
    - Trousers tucked into socks
- You will carry and use repellant during all hours of darkness when you are outside.
- Protect all screens and keep all screened doors securely closed
- Do not stay outdoors all night unnecessarily
- No showers at night unless you go to the shower room clothed.
- Your mosquito bar will be down and tucked in by 1800 hours each day, and during sleeping it will be securely tucked in on all sides. This is one of the most important phases of personal control.

Assam Trucking Company

- Do not remain in your bashas or tents in a state of nakedness. Keep clothes on until you are set to retire.
- To enter the threatre you will be required to possess a bottle of repellant and to be dressed perfectly. [18]

At Jorhat the VD Committee, Civilian Investigation Department, found a number of native bashas operating as brothels, which were visited regularly by enlisted men. A number of cases of venereal disease were reported and treated as a result. In April 1945, the most common offenses committed by the troops stemmed from personnel being caught in Off Limits areas, particularly brothels. The base commander began a more intense educational measure, showing VD films in squadron areas rather than the base theater. Prophylactic kits were distributed to all orderly rooms with a prophylactic station set-up specifically for Squadron E (Black personnel) in its own area. [19]

Trips to Cooch Behar from Lalmanir Hat to visit the Maharajah's palace and elephant farm had to be suspended. Some of the men dined with missionaries, while others frequented more earthly entertainments. The rise of venereal disease was traced back to the brothels in Cooch Behar. The suspension was in force until a more effective method of ensuring that the men would stay away from the brothels was found. Smallpox in the native population around the base prompted wide-scale vaccinations.

At Lalmanir Hat, amoebic dysentery was the main health problem. One carrier of the disease was found among the food handlers. In response to the problem, civilian food handlers were no longer employed in mess halls. G.I. food handlers were required to rinse their hands with a bichloride solution before starting work or when entering or re-entering the mess hall after leaving for whatever reason.[20] Tap drinking water was discontinued and all drinking water was to be taken from Lister Bags, in which water had been treated with chlorine at two parts/million for two hours.

At all bases, sanitary conditions were primarily the concern of the medical community, especially the water sources. Only authorized sources of water were to be used for washing, drinking, and cleansing of mess equipment. The water used for drinking, brushing of teeth and in the mess, had to be boiled, chlorinated or iodized. The source for drinking water was required to be a minimum of 100 feet from a latrine.

By early 1945 a central water system, which had been in the process of being installed for many months, was never completed. The British had been in charge of the project.

Recurrent water system failures were the order of the day, not the exception. At some bases the Army had drilled wells, but no water pumps were issued. The Brits claimed to be only responsible to install hand pumps. The Unit Table of Organization and Equipment (TO&E) had no regular hand pumps listed, so Supply wouldn't issue the pumps. The units were not allowed to have them.

At Chabua, good old Yankee know-how was employed when water pumps were not available for the wells drilled by the Army. A small fuel pump was modified to answer the need. The TO&E was changed shortly thereafter for ATC units.

Only fresh food sources on the approved list were to be used, and had to be inspected by the Veterinary Officer of the Base Surgeon in accordance with Army Regulations. Strict rules in place were as follows:

- Protection of food from insects and rodents
- Prohibition of raw vegetables
- Inspection of food handlers on a regular basis

Assam Trucking Company

The lack of base laundry facilities at the units was responsible for the "dhobi itch," a dermatophyte fungal infection of the groin area and an irritant similar to poison ivy caused by vegetable marker dye. Members of the Indian Dohbi caste did the laundry in an area known as the dhobi ghat. [21] As they washed the clothing, they chewed Betel nuts, which contaminated the clothes being washed. The Betel nut stains on the clothing irritated the skin of the airmen, causing a rash. Dhobis usually did the laundry in large water-filled bins with an adjacent concrete slab. If a relatively "clean" way to wash was not available, any convenient stream or water buffalo wallow was used. As a result, contact-transmittable diseases indigenous to the area were prevalent. The primitive washing method of beating the clothes on a concrete slab or a nearby rock was hard on the clothes and soon reduced the clothing to rags.

In the Letters to the Editor of the Feb. 23rd, 1945, edition of The Dragon, Pfc Don Ballinger asked about a uniform maximum laundry rate. The dhobi Wallas had not established a standard rate. The editor advised Pfc Ballinger that an unnumbered base memorandum stipulated a uniform maximum laundry rate. Anyone paying more should report the abuse to the Provost Marshall.

At Misamari, two washing machines were made from 50-gallon gas drums, salvaged airplane parts and driven by a home-made generator. Before washing the clothes, they were marked with the owner's last initial, plus the last four numbers of his serial number. Once washed and dried, the clothes were ironed by native laborers and bundled. A laundry ticket listing all items was attached to the bundles, which were then brought to another room, where the native laborers were supervised by American soldiers to ensure all bundles were placed in the correct large metal bins and dispatched to the correct Orderly Room for pick-up.

As the seasons changed from winter to spring/summer, the humidity as well as the temperature rose. Colds and heat rash due to poor ventilation were added to the list of other health problems. Ceiling fans were requested but never received. Respiratory, intestinal, and fungal diseases flourished. The living quarters enhanced the development of disease. Excessive sweating and poor ventilation in the bashas accelerated reproduction of flies, bacteria and vegetation. [22]

After many rashes and infections, men began believing the medical officers and complied with established sanitary rules. To rid the latrines of parasitic insects, daily spraying with kerosene began. After spraying the areas, a match would be dropped into each closet to burn out the insects. When the kerosene supply was exhausted, gasoline was used—*once*. The ensuing explosion completely demolished the latrine. There were no casualties.

Chapter 5.0
— Morale

The turmoil at the top of the commands did nothing to help with the morale of the lower ranks. In 1943, Chinese National Aviation Corporation (CNAC) tonnage was greater than that of the Air Transport Command (ATC). A search for the cause of inefficiency and finger-pointing led back to the early days when the India-China Wing-Air Transport Command (ICW-ATC) was still part of the 1st Ferrying Group. Over the protests of Brig. Gen. Robert Olds, who had been involved in the creation and organization of the Air Corps Ferrying Command, the fledgling ICWATC was placed under the control of the Tenth Air Force. Supplies and equipment earmarked for the airlift were often placed in the theater pool, and were issued to other organizations who, according to their commanders, needed the much-prized provisions more than China. Aircraft for the airlift were given other assignments. With the mission in turmoil, bad living conditions, slow promotions, slow mail service, and lack of supplies and replacements, many offered suggestions but little else. Morale, high at the outset, gradually deteriorated, and in the autumn of 1942 reached a dangerously low point.

In December 1942 the organization control was removed from the theater and placed directly under ATC Headquarters in Washington, D.C. General Stilwell retained control of priorities, but operational control was given to Col. E. H. Alexander of ICW. Morale increased, but the sources of discontent remained. Relations between the Tenth Air Force and ATC continued to spiral downward, with each calling the other "robber."

General Stratemeyer reported in May 1943 that morale was bad and efficiency was low among the experienced ATC units, but those newly arrived and inexperienced troop carrier outfits were already in full operation. The fight continued over which units had inviolable rights to supplies being received. After a meeting in 1943 with the CBI commanders, Maj. Gen. Barney M. Giles, Chief of Air Staff, expressed confidence in Col. Alexander.

Capt. E. V. Rickenbacker had a different take. Having visited CBI at the same time as Gen. Stratemeyer in mid-1943, Rickenbacker suggested to the Secretary of War that ICW should again be placed under the control of the theater commander. Maj. Gen. Clayton L. Bissell, commander of the Tenth Air Force, who had a vested interest in obtaining access to the ATC mission supplies and aircraft, agreed. With the ICW coming under Theater control again, Bissell would have the ability to further take from the ICW mission.

Rickenbacker's report listed other reasons for the failure of Hump operations listing in order:

	PROBLEM	CAUSE
1.	Lack of capable and efficient management at the top	*Pugnacious generals, from several different countries, e.g., Generalissimo Chiang, Viceroy Mountbatten,*

Assam Trucking Company

PROBLEM	CAUSE
	General Stilwell (See Fig. 1-5 and Paragraph 2.1)
2. Limited number of airports	British construction, weather, labor problems, etc. Had to be carved out of the jungle and existing tea plantations. (See Paragraphs 4.0 & 4.1)
3. Need of expert communications personnel	No communications systems existed—The system had to be built. (see Paragraph 2.7)
4. Need of expert weather personnel	Nearest weather group in Delhi in the middle of India (See Paragraph 2.5)
5. Need of more radio aids and direction finders	Ft. Hertz only beacon unless the Japanese dropped beacons to lure planes into enemy territory. (See Paragraph 2.7)
6. Need of more qualified engineering officers and maintenance men	Training them as fast as could be accomplished—there was a war after all. Most assets were being sent to Europe. (See Chapter 3)
7. Inexperience of pilots	There was a war on. Most assets were being sent to Europe. (See Chapter 3)
8. Vast difference in pay between the Army and CNAC	Noted by Stilwell in 1942. (See Paragraph 1.3)

Table 5-1, Rickenbacker Report Findings

Rickenbacker also observed that from 1 January to 1 June 1943, 26 ICW transports had been damaged beyond repair, some having collided with aircraft parked on landing strips due to lack of dispersal areas. (See #2 above). What Capt. Rickenbacker had mentioned were known difficulties. Transferring the ICW back to theater command control would only add to the problems, not fix them. [1] For all of his experience as a pilot in WWI, Rickenbacker didn't seem to understand that a new command was being built "on the fly." It didn't look like Europe, but it appears by his comments that the CBI Theater should look and act like the European Theater. They couldn't, however, fit a square peg into a round hole.

Half a world away from their homes and families, living in tents or bamboo huts perched at the edge of jungle, tea plantation and rice paddies, the men of ATC kept flying despite the varied viewpoints at higher command levels. They still had a job to do, with high expectations from all above them. Recreation facilities and programs were lacking causing the spirit of "all work and no play" to worsen. It is no surprise that in early 1944 morale had dropped to an all-time low.

Upon his arrival in 1944, General Tunner remarked that morale was low which, in his experience, contributed to accidents. What he saw in CBI in most cases "were washouts, total losses with planes either flying into mountain peaks or going down in the jungles. In many cases in which there was reason to believe that some or all the crewmembers had been able to parachute from their planes, those men were never seen again. The jungle had simply swallowed them up." [2] The morale issue seemed to him to be even worse among the ground crews. He saw a hopelessness to their effort. There was some truth to his supposition. The work was as monotonous and predictable as the food and the weather.

Assam Trucking Company

A morale board was established and met twice per month. Some of the changes recommended included issuance of sheets to those sleeping on blankets, better food in the mess hall, a rotation points policy, and other complaints heard and acted upon.

5.1 DAY IN THE LIFE

5.1.1 Ground Personnel

For those working on the in maintenance, the typical day started at 0630. After making his way out of his mosquito netting, the load supervisor walked the 100 yards through the mud to the washroom. The washroom had 2 rows of faucets over trough-like sinks made of split gasoline drums. Salvaged fuel drums were also used to make water heaters.[3] At the mess hall sat an Indian paper boy with the Hindustani Standard Statesman, which was published two days per week. The load supervisor sat down to a breakfast of prunes, cereal, dehydrated eggs, grapefruit juice and coffee.

After breakfast he returned to his tent, rolled up the netting, swept the floor, then caught the shuttle bus to the work area. His job was to supervise a group of native laborers as they loaded a truck that would be taken to a waiting plane. He watched the group closely, with the plane manifest in hand. At 1130 he returned to the living area for mail call, then headed over to the mess hall for a lunch of spam, dehydrated potatoes, canned vegetables, fruit salad, coffee, and water.

The afternoon routine was the same as the morning. At 1730 he was relieved by the evening shift supervisor. Every other week the schedule was changed. Instead of the morning routine, he would take the evening shift for a two-week period. He checked mail call again and headed for a dinner of

C-rations, spinach, canned tomatoes, pineapple, coffee, and water. He took a shower, returned to his tent, adjusted the mosquito netting around his bed and set off a bug bomb.

His evening might include a trip to the dayroom, where he would write letters home as he listened to the news on the dayroom radio, engaged in a group discussion of the day's events, or viewed a movie, after which he returned to his tent for the night.

5.1.2 Flight Crews

The flight crews had a 36-hour cycle. A duty pilot would be awakened at 0500 by the Charge of Quarters (CQ). The pilot already knew his mission for the day. He dressed in the dark using a flashlight, and made sure he had his gun and survival knife. He ate alone, then caught the crew shuttle to base operations. The radio operator secured the briefings while the copilot made out the flight plan and the pilot received the manifest and flight clearance. Sometimes a mission assignment would be to Myitkynia to evacuate hospital patients, both ambulatory and on litters. The mission called for them to be in the air by 0700 for the 1.5-hour flight from Chabua to Myitkynia, Burma, for Medevac. The litter patients were loaded first, then the ambulatory patients were assigned seats. Many of the patients were wounded from action received as a member of Merrill's Marauders' (U.S.) or Wingate's Chindit (British) forces fighting the Japanese in the Burmese jungles. In less than an hour, the plane was back in the air to Ledo where the patients were unloaded. Upon return to Chabua, the crew headed for the dispensary for their two ounces of combat whiskey. In 36-hours they would be in the air again. [4]

Assam Trucking Company

If the mission was to Kunming, the crew faced a five-hour eastbound flight, which was dependent on weather, with an average of 2.5-hours on the ground while the plane was unloaded and readied for the return flight to an Assam Valley base. After an 11- to 12-hour day, the crew still had to debrief on weather, communication conditions, any new crash sites not on the maps and any other information the next crew might need. And so, it went.

Lower ranks didn't have the same experience that Tunner described. They found that the work rotations kept them busy, with little time to actually consider their plight. Flight Officer Paul Schaffer's fondest memory of his time on the Hump was that of looking forward to the adventure. He saw a lot of things he would never have seen without the opportunity. His worst memory was of the cold at altitude. His duty was to deliver gas, bombs or whatever to China. He made 65 roundtrips over the Hump in a C-109, known for its proclivity to explode due to gas leakage from the fuel tanks of the B-24 tanker conversion and sparks from the radio. [5]

Gordon Leonard, an experienced pilot with 2200 hours of flight time, became a CBI duty pilot stationed at Jorhat. After two to three flights, he was considered sufficiently qualified to be a first pilot; two to three more flights and he was considered a check pilot. The routine in the command was for each pilot to have a check ride every 90 days to ensure their performance was still acceptable and within command requirements. From that point Leonard made most of his flights in the right seat, evaluating pilots and completing check rides. "All we did was fly, eat, sleep, gamble a little, drink a little. There was no time for tours." He just wanted to complete his time so he could return to his family in the States.

Leonard detected little "battle fatigue." There was a story about one man who dropped down to 80 pounds, yet refused to have the flight surgeon ground him. All he wanted was to get his points in and go home. He had to carry a pillow to sit on because his pelvic bones were breaking through his skin. He looked like a Prisoner of War (POW) but managed to complete his hours and return home. Leonard never felt that his morale was affected.

The thing Leonard missed most was something cold to drink. Rupert's Beer was in abundance, but basically tasted lousy. Packed in cardboard cartons and sawdust, 1 in 5 or 1 in 4 bottles were spoiled, probably because the cartons had sat on a dock somewhere in the heat. Each officer was given a case of beer per month. Beyond the spoilage, cooling the beer was difficult, but Yankee ingenuity won out. Enlisted maintenance personnel would wrap a few bottles in a burlap bag dipped in gasoline and twirl it over their heads. The rapid evaporation (if it wasn't raining) would cool the beer. On occasion before a flight took off across the Hump, a sergeant or an airman would bring a case of beer to put on board. By the time the crew got back across the Hump, the beer would be cold, if not frozen. The owner of the beer would be waiting at the revetment when the plane landed. Sometimes an aircrew member would be asked to take a canteen filled with water so the canteen's owner could get a cold drink. The Officers' Club always had gin, which some felt helped. Still, there was no ice in the mess halls for the water pitchers, which sat in the middle of each table. Living in a warm, damp climate seemed to always give one a thirst. [6]

5.2 ENTERTAINMENT

Entertainment throughout CBI was limited. Reading materials were scarce; movie projectors were few in number and in constant need of repair, as with everything else. Despite the projector problems, three films were received at Chabua per week and distributed to various areas on base. G.I. shows were organized by men with professional experience. A weekly variety show was performed on bamboo stages. USO shows with performers like Keenan Wynn, Joe E. Brown, Pat O'Brien and Jinx Falkenburg found their way to CBI. There was negative feedback on the USO shows. It appeared that the USO shows had created hostility because of the attitude of the show people. There had been a problem with scheduled shows arriving in theater.

Baseball experts Dixie Walker, Luke Sewell, Paul "Pop" Warner, and Arthur Patterson made the rounds of the bases.

Lalmanir Hat had been missed by most of the shows presenting top Hollywood stars, who usually stopped at Lalmanir Hat on their way to Assam and China, but not to present a show. While other bases saw the USO Show "Happy Holidays" with Sammy Cohen and a troop of five men and five women, Lalmanir Hat received the USO troupe #269 with five women, who arrived on December 11, 1944. Their presentation was anything but classical, skirting the domain of vulgarity.

As a substitute for the expected USO show, the Red Cross Clubhouse at Lalmanir Hat had an open house for both enlisted men and officers for Christmas, complete with carolers from the USO, a church service and turkey dinner.

Fifty-four special Christmas boxes contributed by Mrs. J. L. Cruze, whose son was based at an ICD base in India, were distributed. Chaplin Unger, by arrangement with Search and

Rescue, secured a C-47 and played "Aerial Santa Claus," dropping Christmas gifts over remote outposts in North Burma and India by parapack.[7]

5.2.1 RADIO

The 16th AFR Station VU2ZK started broadcasting May 27, 1945. The programing for the station started at 0600 and ended at 2300 with a touch of home.:

- 0600 Musical Clock
- 0655 News
- 0700 Personal Album
- 0715 Sammy Kaye
- 1130 TBA
- 1145 Melody Roundup
- 1200 Music U.S. Loves
- 1230 GI Jim
- 1245 News
- 1300 Requests
- 1700 Words with Music
- 1715 Spotlite Bands
- 1730 Jill's Juke Box
- 1800 Gildersleeve
- 1830 GI Journal
- 1900 Music Hall
- 1930 News
- 1945 Johnny Mercer
- 2000 Kay Kyser's College
- 2030 Frank Morgan
- 2100 Music We Love
- 2130 Nelson Eddy
- 2200 Globe Theater

- 2230 One Night Stand
- 2300 Sign Off

Despite the radio station, the main form of entertainment remained the movies.

5.3 ACTIVITIES

1945 brought a number of changes to the Assam Valley bases. Flights and tonnage increased, personnel numbers went up with the European draw-down, and amenities increased.

Church services for Protestant, Catholic and Jewish personnel were well-attended. When the assigned Protestant and Catholic Chaplin rotated to the states, Protestant services were conducted by Rev. Reuben Holms of the Jorhat Baptist Mission School. Chaplin Hayes of the 1333rd AAFBU held services for the Catholics.

A Chapel Club with over 75 members had a number of activities featuring guest speakers and banquets. A group of three from the Chapel Club visited local missionaries to the Boro people, one of the Hill tribes. To reach the mission, the three took a train, motorbus, bullock cart, elephant, then did a little walking. One village had never seen Americans, so the village leaders declared a school holiday for the occasion.

Sporting events comprised boxing bouts, softball and volleyball—unless there was rain. The boxing team went to Calcutta for a tournament.

Special Services activities increased at Mohanbari in September 1945. Hunting and fishing trips were organized. The stage shows "Who Was That Woman" and "Swing Lively" were most-liked by personnel.

A small gym and recreation hall were built at Lalmanir Hat in 1945. A 12-team basketball league was organized as well as six softball teams. The 32nd Malaria Control team was declared the winner of the basketball league after winning ten straight games. The highlight of the league was the banquet for the top three teams, who feasted on steak.

"Outdoor movies were shown in a theater without a roof. Requests for funds were turned down. The base was trying to find a way to locally procure materials to repair the roof before monsoon season." There were no movies, libraries, or towns close to any of the Assam bases.

An education program, known as the "Bamboo University," had one hundred and eighteen men enrolled. With the end of the war in site, topics for discussion included:
- "Is Communism a Threat to America?'
- "What Should be America's Post-War Armies?"
- "What Shall We do with the Axis War Criminals?"

The War Room map was very popular as a way to follow the war in Europe and current events as the war appeared to be drawing to a close. [8]

At Misamari a new oven had been installed, which turned out fresh rolls and pastries. New ventilators were installed in the Line Mess, and iced drinks were served at least once a day *if* the ice plant was working. Ice cream and ice were available at Mohanbari, Misamari, and Jorhat. Food supplies were better as was the water supply and electrical power. Plus there were no QM shortages. [9]

In January 1945 the Public Relations Section at Jorhat conducted two tours for 61 civilian allies—British tea planters and Indian students. The tour groups were briefed on the Operational and Maintenance functions of the base then were shown the film "The Mission of the Air Transport Command." The tours promoted better relations with the U.S. Allies. At the same time, 62 Air Medals for 150 hours of mission flying

Assam Trucking Company

plus 29 Distinguished Flying Crosses for 300 hours of combat flying hours were awarded at Jorhat.

With the flights settling into a routine, the various base commanders stepped up activities to improve morale. At Jorhat the first edition of the "Hump Express," an in-theater newspaper, was published; and American, British, and Indian troops played exhibition matches in soccer, softball, volleyball, rugby, and tennis. Athletic facilities built included:

- One archery area
- Fourteen badminton courts
- Three softball fields
- Two basketball courts
- Ten horseshoe courts
- Fifteen volleyball courts
- At one track area a Volleyball League was organized in addition to tennis and badminton tournaments and a Squadron Sports Carnival. [10]

A hobby shop where personnel could do leathercraft, water and oil painting, drawing, clay modeling and sculpting was put together by Cpl. Frank E. Stafford and Sgt. Robert Thurman from the Utilities Section at Misamari in April 1945. That month the Public Relations section submitted 44 stories to the theater publications the *Hump Express* (official newspaper of the India-China Division, Air Transport Command) and the *CBI Roundup*. [11]

At Chabua in March of 1945, Renshaw University, an Information and Education school, had approximately 200 enrollees and offered Courses in the following:

- Bookkeeping
- Algebra
- Journalism
- English
- Music Theory

- Business Law
- Arithmetic
- Typing
- Sketching and Design

Some of the classes offered were suspended when rotation and transfers reduced the student and instructor populations.[12]

A radio show was originated, written, and produced by Pfc R. Vern Beckwith, titled "How to be a Civilian." It answered questions about G.I. Bill rights and highlighted colleges. [13]

5.4 ROTATION POLICY

The surrender of Germany was a harbinger of things to come. Many felt their time to go home was close at hand. It seemed like a positive event, but turned out to be a negative for the men of ATC in CBI. Since most of the time in CBI was not considered combat time, credits for rotation to the States were hard come by. While flying hours spent in a slow-moving, unarmed, unescorted transport were considered combat conditions and counted toward medals, they did not count toward rotation credits. A reduction in the critical score was anticipated by summer, with the hopes of more ATC men being redeployed and eventually discharged. Instead, the rotation system was suspended. All available planes and ships were needed in Europe following V-E Day. Operations in CBI were to be stepped up to put pressure on in the Pacific and China.

Due to the rotational requirements, a number of applications were received for 45-day temporary duty assignments stateside. They got a break with the cessation of hostilities in Europe. The number of applications fell off. [14]

Assam Trucking Company

In March 1945 at Lalmanir Hat, with the war turning for the better in Europe, morale appeared higher. There was increased incentive to continue. The effects of the new rotation policy were being felt. [15]

"There had been no definite rotation program in place before" thus giving the men little hope of coming home before the war ended. When Tunner arrived in CBI, he found the rotation plan was dependent on the number of flight hours (650). Some had been flying as much as 165 hours per month to get their hours logged within 4 months. The concept was to get the required flying time done so they could go home. The missing piece was someone to take their place. There were none.

Tunner felt the existing policy was dangerous and had the rotation plan reworked. Once revised, the fliers' point requirement had gone to 750 hours *plus* they had to have 1 year in theater [16] Adding to their misery, the aircrews learned they were limited to 100 flying hours per month. The rational in lowering the number of hours per month was to prevent accidents. This was not a popular change, but it was the decision of the commanding general.

For others, the rotation plan was different. Having the right number of points didn't guarantee a trip home. The owning unit had to have authorization to release an individual. For example, one unit with 18 enlisted men eligible for rotation could release only as many as authorized. The authorization may be for only 1 person. A unit requirements review would be made of the 18 eligible men and, based on points and availability of Military Occupational Skill (MOS)-skill set, the selection of one would be made. All 18 men had points ranging from 85 to 104 toward the number of points required.[17]. In accordance with the review, the one with the required 85 points, but with the least points, could in theory rotate home first, leaving the other 17 still in theater. It came down to which skill-set (MOS) was most needed, and who could be released from duties without a negative impact to the unit.

By August 1945, stateside rotations started with enlisted men who had left the U.S. prior to 13 June 1943. They were to be rotated home by the end of August. The next group included enlisted men who had left the U.S. Between 13 June 1943 and 25 July 1943. Orders were issued by Division for all men with 85 or more Adjusted Service Rating points calculated as follows:

- Months in service overseas = 1 point/month
- Combat awards (including campaign medals and battle stars) = 5 pts/medal
- Dependent child under 18 = 12 pts/child

Men over age 40 were processed for discharge. [18]

5.5 PERSONNEL

Many of the enlisted ground crews had 850 to 1,000 duty hours with no possibility of rotation in sight. The time in theater for flight crews was almost equal to that of ground personnel. Officers were allowed to rotate upon completion of hours required, but not the enlisted because of the need for specific skill sets. Nearly 50% of R/O's were Pvts or Pfcs. Between January and August 1943 not one was promoted, while others in less hazardous MOSs were receiving promotions. To alleviate the promotion problems, the men who had completed their hours for rotation were given ground jobs, e.g., admin clerks, flight-schedulers, or flight check instructors. The actions in Europe, Africa, and the South Pacific, plus support of the war effort at home, had depleted the number of available men to replace those in CBI.

Assam Trucking Company

It didn't help that CBI was basically the forgotten theater of the war.

Project 7 was part of FDR's promise to give more aid to China in the form of planes and aircrew who were not trained. One perception existed at Tezpur that part of their personnel shortage would be alleviated. When the personnel and planes assigned to Project 7 arrived, that perception was crushed. Project 7 turned out to be a number of aircrews assigned to C-87s which they flew from Natal, Brazil. They arrived at Tezpur on August 1, 1943. Their ranks were a point of contention, since most of them out-ranked those at Tezpur, who had not had a promotion in 12 months. Morale issues were further increased when the anticipated shorter hours with the influx of personnel did not materialize. [19]

At Jorhat in January 1945, there were 309 officers and 2,040 enlisted permanently assigned, with 26 officers and 324 enlisted on attached duty. Yet there was a personnel shortage with mal-assignments evident. On base were the U.S. Navy O_2 Unit, the Army Post Office, Malaria Control, Signal Service, Weather, Military Police, Aviation Engineering, and Army Airways Communications. The Intelligence and Security Section was responsible for briefing of flight personnel on the routes—terrain, native inhabitants, customs, and escape information. Ground personnel were in charge of security, handling of classified materials, censorship, camera registration and currency limitation.

Another glitch popped up when new radio operators who were veterans of the European Theater were assigned to CBI. There was tension at first because the veteran group were T/Sgts while the R/O in CBI were Pvts to Sgts. The tension eased when the group in CBI realized that the new group coming in were now serving a second tour of duty having endured flak and enemy attack over Europe.[20]

In April 1945, Misamari, received new arrivals, formerly R/O crewmembers from the 8th AF in Great Britain, who found themselves now assigned to the Communications section.

5.5.1 Women

Based on the experience of nurses and Red Cross women already in the area, the deployment of WACs in CBI was not encouraging; nurses had previously suffered from clothing shortages, emotional strain, tension, and a medical evacuation rate four times that of the men. Many Army officials believed that women would not be able to stand for long the climate and diseases found in Southeast Asia. Earlier attempts to bring in WACs had been blocked by Lt. Gen. Stilwell for two years. He yielded to the requests from the Air Forces' commander Maj. Gen. George E. Stratemeyer only when Stilwell received the assurance that any WACs brought in by the AAF would never under any circumstances be assigned to other than Air Forces headquarters. The first WACs assigned to CBI were posted to Headquarters in Calcutta in July 1944. December 1944, Major Betty Clague, Staff Director of the WAC, briefly surveyed the base at Lalmanir Hat to decide its suitability for WAC personnel.

After careful preparations and attention to the needs of the women in theater, 300 enlisted were assigned to fill stenographer, typists and other highly skilled MOSs. "Never has any WAC contingent received a more cordial welcome than this group," wrote the WAC staff director, Maj. Betty Clague whose later review of facilities at Lalmanir Hat stated that Lalmanir Hat did not meet the requirements originally established. Lalmanir Hat was more remote than others in the Assam Valley and was not a headquarters base. [21]

5.5.2 Blacks

To help with the lack of personnel, Blacks were added to the mix, with some trepidation. At Misamari, over 400 Blacks reported for duty in October 1944. At the time they arrived, morale had been high. There was some uncertainty as to how the white personnel would accept the Black groups. Because of its remote location, Misamari was self-contained and insulated. It was felt that what would be a small infraction or disturbance in a state-side base could, if it were to occur at Misamari, have an exaggerated effect.

When Maj. Vern Clements, Jr., Director of Personnel and Administrative Services, interviewed Lt. Col. Claron Pratt, Base Commander, the colonel reported that he had less difficulty than he had anticipated, especially on the social side. Jim Crow was alive and well and living in CBI. Separate but equal facilities were provided in the Black Area V, e.g., theater, PX, etc., yet the Black soldiers were allowed to patronize the facilities available for the white troops. As a result, no friction was noted. The Base and Area V Theaters were used for religious services on Sunday and Wednesdays. Daily Catholic mass was conducted in the Base Theater.

The February report for Misamari related that the transfer of white personnel from the base caused a shortage of qualified men in a number of departments. It was decided to solve the shortage through reassignment of Black personnel. The result of the effort was that, in view of what was considered their insufficient aptitude and ability to acquire a higher degree of skill, the Black troops couldn't be generally considered as potential substitutions for skilled white personnel. Therefore, the best use of Black personnel was as plane loaders, drivers and as basic duty soldiers.

Just a month later, some of the supervisors' comments exhibited surprise at the capabilities of the Blacks. Capt. Sanford Agnew, Director of Priorities and Traffic, stated that he had used as many as 128 Blacks in his department at one time and found they exhibited the same degree of skill shown by white men in the particular job assigned. Used originally as plane loaders, Black personnel were appointed to supervise the Indian loading crews when Indian labor was used to load planes. Capt. Agnew felt that the Black personnel needed a little more direct supervision and someone to listen to their problems more-so than the white troops.

Efforts were being made to improve the status and morale of the Black personnel. One Black man was working as Assistant Supervisor. It was expected that in the future more Blacks would be working in office work and more responsible duty areas. The Priorities and Traffic training course would make it possible for those taking the course to receive a semi-skilled rating, qualifying them for upgrade classifications. It was felt that the Blacks believed they were being given a chance—an even break with white personnel. [22]

5.6 CRIME

There were eight Summary Courts Martial, with the most common offense being a violation of 96th Article of war, discharging a service pistol in a public place. There were 3 Special Courts Martial for smuggling and sale of cigarettes in China.

Other offenses punishable under the 104th Article of War were for various minor offenses such as traffic violations, lateness, improper dress or improper conduct at parades.

Additional disciplinary problems apparently unique to the Black troops included malingering and the petitioning of

Assam Trucking Company

higher echelons for redress of imagined grievances without regard to official channels of correspondence. Consequently, an excessive number of requests for discharge were made, based on claims of dependency, undue hardship, physical disability or some other naïve ground. It was felt that, although the courts martial rate was proportionately higher for the Black troops than the White troops, the commanders involved saw no reason for alarm.

BLACK TROOPS				WHITE TROOPS			
No.	A of W	Type	Month	No.	A of W	Type	Month
1	93rd	General	Jan	2	93rd	Special	Jan
1	96th	Special	Jan	1	96th	Summary	Feb
2	96th	Summary	Feb	1	96th	General	Feb
2	96th	Summary	Mar	4	96th	Summary	Mar
3	96th	Summary	Apr	2	96th	Summary	Apr
1	96th	Special	Apr				
10	TOTAL			10	TOTAL		

Table 5-2, Courts Martial Comparison at Misamari

It was noted by the Provost Marshall Lt. John J. Zemerlin, who received the Black troops, that most of them were very responsive to the rehabilitation programs, resulting in a zero rate of recidivism. Misamari had a different approach with programs and activities and the availability of base facilities, e.g., PX, Theater, etc., to the morale of the Black troops, which was credited with the lower rate of crime among them. [23]

Unlike the experience the Blacks had at Misamari, others in the Theater would not be treated well. The discrimination and ill-treatment often led others to find avenues of escape through opium and gangja (Marijuana), which were readily available in the area. The ill-treatment and the drugs led to the one murder reported of a white officer by a Black soldier. Pvt. Herman Perry was a member of the 849th Engineer Aviation Battalion, a Black unit commanded by white officers. The 849th was one of the units sent to India to build the Ledo Road. Having used drugs the night before, Perry was reported AWOL on March 3, 1944, after missing reveille. He had spent time in the stockade and knew he did not want to return there because of the harsh treatment he had received. Perry resisted arrest but was not disarmed. He managed to escape and hitch a ride to Nagaland, home to the Naga headhunter tribe. This lack of attention to detail cost Lt. Harold Cady, who was known to Perry as a hard disciplinarian, his life. Cady was aware of the arrest warrant on Perry and attempted to take him back into custody when he was murdered.

In addition to murder, Perry was also charged with desertion (also punishable by death) and several counts of willful disobedience. When he was finally arrested and taken into custody, he was tried and convicted by an all-white panel and sentenced to death. He was executed March 15, 1945. [24]

In June 1945, five men were in the guardhouse at Lalmanir Hat: one for being AWOL, three for being off base without passes, and one for setting fire to the latrine while drunk. A member of the 308th Bomb Group was under investigation by Intelligence and Security at Rupsi, Burma, in August for allegedly making slanderous remarks about China and the Generalissimo. The investigation had been initiated by the Generalissimo's headquarters through General Wedemeyer

Assam Trucking Company

With the war over in Europe and the movement of the MPs at Lalmanir Hat to Shillong, there was a rise of petty theft among the locals. The Indian nationals were trying to get their hands on something before the Americans left. With the upcoming elections, the town of Lalmanir Hat and the surrounding area saw an increase in political activity among the Congress Party, the Communist Party, and the Muslim League. All known party members were kept under surveillance.

Local Bengal police were employed to assist the Intelligence and Security Office and the Provost Marshall to curb incidents with Indian civilians. This continued until the local laborers complained about being "shaken down" by Bengal police. When the U. S. authorities heard the complaints, the corrupt plainclothes policemen were discharged. The entire force was affected. The Chief Clerk of the Civilian Personnel Office was implicated, however there was insufficient evidence to prosecute. There was enough evidence for the U. S. authorities to discharge the Clerk and bar him from ever being employed by the U. S. government again. [25]

5.7 PHYSICAL CHANGES

The PX building at Misamari was painted white inside and out, and the terminal was painted Black. Inside the exchange was decorated with brightly colored Black Vargas and Macon girls in chalk. The Snack Bar was near completion. The planned menu included hamburgers, ice cream, cold fruit juice, coffee and doughnuts. They hoped to add Coca-Colas, sundaes and milk shakes at a later point.

In December 1944 a new firehouse was under construction. By March of 1945 the new Base Library had over 3,000 books. To further boost morale, a Base Sweetheart contest was held. The winner was Miss Ann Elizabeth Taylor of Columbia, S.C. Her sponsor was her brother, Sgt. John H. Taylor, alert crew chief. She was sent a silver loving cup. The war ended before a second contest was held.

Mohanbari unit strength in September 1945 totaled 1993 —1202 white enlisted, 349 Black enlisted plus 442 officers (375 flying and 67 ground personnel). There were no mal-assignments noted in the report. [26] The end of operations was near.

Chapter 6.0 — Planes

At the time of the attack on Pearl Harbor, few medium and long-range transports existed in the US Army Air Corps arsenal. Only eleven converted B-24s and 40 to 50 twin-engine planes belonging to the 50th Transport Wing represented the transport aircraft available to the Army Air Corps (AAC). The AAC was not prepared to support any variety of airlift, especially not one of the magnitude required by the Hump. Per executive order, the civilian airlines were compelled to turn over a portion of their twin-engine transports to support the war effort. Of the 209 transports operated by the airlines, 89 went over to the USAAC. Eighty-nine transports would not begin to meet the airlift need.[1] The number of aircraft thought necessary to support the airlift was based on the operations of cargo planes over the hump in the months prior to September 1942—the only data available.[2]

The aircraft manufacturers were producing bombers and fighters, not transports. However, the AAC had made a large procurement commitment before Pearl Harbor for DC-3s (C-47s) and DC-4s (C-54s). The producer's response to deliver aircraft quickly resulted in more orders from the AAC.

However, deliveries proved to be problematic. The aircraft as used by the airlines needed cargo floor and door modifications to support the weight and configuration of the cargo to be airlifted.

As the Air Transport Command (ATC) took over control of the airlift of supplies for China, other aircraft were brought into the arsenal. Introduced to the Hump in 1943, the C-46 Curtiss Commando answered the need for increased cargo tonnage. Able to carry almost twice as much as the C-47, the C-46 seemed to be the answer to raising the tonnage figures. Two other planes, the C-54 (DC-4) and the C-87 (B-24 cargo conversion) were introduced at the same time. The C-109 was another B-24 conversion used as a tanker, essentially as a fuel bladder, to haul aviation fuel in larger quantities. Unfortunately, as with anything else, each of these aircraft had its own set of problems.

6.1 C-47, Skytrain

With the fall of Burma to the Japanese as far north as Myitkyina in 1942-43, Hump flights had to take a more northerly route over higher elevations. Of the available transports, the C-47 had a service ceiling of 23,500 with a full cargo load of 6,000 pounds. The China National Aviation Corporation (CNAC) started the cargo flights with the C-47 Skytrain, (DC-3). Also known as a "Gooneybird" or Dakota (British conversion), the C-47 became the workhorse of the Hump operations.

Because of the wing configuration as a "low-wing monoplane," the fuselage of the C-47 stood too high off the ground, making loading from a regular truck platform impractical. Modifications made included:
- Installing a regular cargo door

Assam Trucking Company

- Reinforcing the cargo floor
- Developing special cargo loading equipment. [3]

At some bases, elephants loaded barrels of aviation gasoline and other cargo. In May 1942 President Roosevelt directed the Secretary of War to commandeer all C-47 transports flown by civil airlines in excess of 200 and to refit them for use as transports in support of the war effort. [4] The C-47, as used by the commercial airlines, was readily available in large numbers. Production continued through the war until, by the end of the war, over 10,000 C-47s were in use.

6.1.1 Technical Specifications

Engines:	Two Pratt & Whitney R-1830 Twin Wasp radials, 1200 hp each
Armament:	None
Speed:	230 mph at 8,500 feet
Climb:	1,130 feet/minute
Ceiling:	23,200 feet
Range:	1200 miles
Weight	
Empty:	16,865
Loaded:	30,000
Wing Span:	95 feet
Length:	64 feet 6 inches

6.2 C-46, CURTISS COMMANDO

Originally introduced as a competitor to the DC-3, the C-46 was designed in 1937 by Curtiss-Wright in St. Louis, MO, as a high-altitude, 36-passenger commercial transport, but it didn't enter full production until after the attack on Pearl Harbor. During 200 hours of testing, Eastern Airlines

reported that the C-46 carried 10,000 pounds of cargo at an average indicated airspeed of 200 MPH and consumed 135 gallons of fuel/hour. Untested in a military setting, the C-46 had a service ceiling of 25,000 feet with a cargo load of 10,000 pounds. Unfortunately, the plane's performance, while excellent, did not hold up for trips over the Hump. The first three delivered to Mohanbari had hydraulic problems and crashed. [5] Less than encouraging reports started coming in, especially when viewed from the environment in which the aircraft were to be flown and included:
- Leaking fuselage in heavy rain.
- Peeling camouflage paint
- Malfunctioning hydraulic and fuel system

In August 1942, fifty-three immediate modifications were requested, not including winterization—an amazing note in light of the environment in which the aircraft would fly. An additional 46 modifications were labelled as desirable. The first thirty C-46s delivered to ATC were sent back for modifications. Pilots wanted 4-engine aircraft instead of the less than reliable C-46. With no four-engine aircraft available, and offering twice the value of the C-47 and more efficiency than the C-54 over distances of 1,500 miles or less, the C-46s had to stay. Assam became a huge experimental laboratory for the Curtiss-Wright Commando. [6]

By the end of 1943, 363 C-46s were accepted, plus another 1,321 during 1944. By the end of the war a total of 3,123 had been accepted for use by the War Department. These numbers were staggering when the following drawbacks were considered:
- By November 1943, Curtiss-Wright reported 721 production aircraft required changes
- Maintenance crews labelled the plane a "plumber's nightmare."

Assam Trucking Company

- From May 1943 to March 1945, ATC received reports of 31 C-46s either catching on fire or exploding in flight. The gasoline heaters built by Curtiss Wright were believed to be the cause of the mid-air explosions. Heaters were changed to a Janitrol brand, which used the avgas from the plane.[8]
- Others were thought to have crashed due to vapor lock (hot weather), carburetor icing (altitude) or other defects. Vapor locks caused the engines to temporarily cut out in flight or on takeoff. The result on takeoff was the loss of a crew and a plane.
    o One modification made was the installation of a submerged centrifugal fuel pump in the wing fuel tank. The worst maintenance feature was the riveted construction of the wings of the early models. Wings had to be removed to replace the fuel tanks, which took an average of two days down time (150-man hours).
    o Carburetor icing caused when ice formed, blocking the entrance of the air induction system. The two methods identified to remedy the situation were
- Pre-heating the air before introducing it to the carburetor
- Placing alcohol jets just before the carburetor. In theory, the alcohol was supposed to "dissolve" the ice because it has a lower freezing point. The trick was to find the most effective method: heat alone, alcohol alone or a mixture of heat and alcohol. The best method turned out to be the introduction of heat just before entering a heavy moisture atmosphere. Alcohol alone aggravated the icing

situation. In mid-1944 the alcohol carburetor anti-icing system was disconnected on all C-46s flying the Hump.
- Fire extinguishing system—Lines were the most difficult thing for a mechanic to replace
- Hydraulic booster system. The main system provided 1200 PSI to actuate the landing gear and flaps, while the auxiliary system provided 800 PSI as a booster of the controls. The hydraulic dump valve would stick and begin to smell. The crew chief had to open the trap door to access the valve and unseat it by beating on it with a hammer.
- Hydraulic Prop—Tendency to go wild. Most down time in China was due to props.
- Flight controls. There was difficulty in synchronization of throttle settings, worn bearings and corroded shafts.
- Push rod oil leaks. [9]

According to Herbert O. Fisher, former Curtiss Engineering test pilot and technical representative, the C-46 was the biggest problem the maintenance crews had to face. The Curtiss C-46, "was aerodynamically sound and structurally a Goliath. However, there were many inherent problems. As a sample of some of the items and system malfunctions, fuel and hydraulic systems, carburetor shrouds and deicing ability, flight instruments, ignition, cockpit lighting, wing landing lights, cabin heaters, fire bottles, Dzus fasteners, inadequate engine cooling, selectors bypassing fuel from tank-to-tank, batteries, generators, voltage regulators, fire extinguishers, warning lights, quality of hose connections, auxiliary power unit failures, and *many* other unsatisfactory items. Fisher further added that the single most common cause of crashes and bail-outs was due to engine failure and

Assam Trucking Company

carburetor icing. Airmen were not provided with printed materials or directives on the proper use or application of carburetor heat." [10]

The two-engine aircraft, first flown in 1940, was put immediately into war-zone utilization in 1941 without extensive service experience on all its systems. The demand in the Africa, Pacific and CBI theaters of war for transports made the implementation necessary. Herb Fisher spent 14 months overseas in North and Central Africa, the Middle East and CBI, making 96 research missions over the Hump. [11]

Training on the C-46 was another issue. Most new crews arriving in India for assignment to the Hump run were not prepared for the situation in which they found themselves. Most were considered "ninety-day wonders," hurried through flight schools in the United States, then sent on to an Operational Training Unit (OTU) specialized base for "further training." These young men, most in their early twenties, were then sent overseas to serve as copilots over the Hump. When queried about being "checked out" for C-46 service, one "Humpster" chuckled and said, "No, I had two take-offs and two let-downs at Reno, NV, then found myself in India. We were not prepared for what we found." Most were imbued with a feeling of adventure yet few realized the very real dangers they faced. [11]

When the first C-46 arrived at Chabua, there were no spare parts, technical manuals or trained maintenance personnel. In addition, the new plane had not been completely service-tested in the States. With the emphasis placed on planes and not on the needed parts, by the middle of 1943, the parts shortage had become critical. There was a lack of consumption figures—nothing to tell the Air Service Command (ASC) what was needed in the field by ATC. No consumption records led to the disinclination of the Air

Service Command to accept ATC utilization figures. The antagonistic attitude between ATC and ASC was caused by a mutual lack of understanding of each other's problems. Maintenance equipment and tool shortages remained critical throughout 1943. "Deals" to acquire such equipment as air compressors, jacks and engine hoists became the practice of the day. Letters to higher headquarters in the states brought no relief. [12]

Engine fires, a major problem, would cause the wings to melt. The only thing to do was get out of the plane! [13] Accidents occurring immediately after takeoff were fatal in a number of instances. Some pilots flew the planes through trees, shearing off the wings and bending props. Others managed to circle the field and land safely, fully loaded.

Pilot error was the major contributing factor to a number of C-46 accidents. The plane was powerful and but not fully tested in theater. The aircrews made mistakes because of a lack of documentation (including checklists), truncated training, and lack of time in the aircraft. Instinctively, based on their previous experience in other planes, pilots of a C-46 with a capacity load had a tendency to reduce power too soon after takeoff. Lowering the nose to maintain or gain air speed actually had the opposite effect, resulting in the loss of altitude. At night, the landing lights were turned off immediately after takeoff. The crew was unable to see the proximity of the plane to the ground, often leading to a critical error in total blackness. If not set properly at night, a lack of proper cross-check between the artificial horizon and air speed indicator resulted in disaster.

Pilots often did not apply power rapidly on initial power application, resulting in taking up too much runway before sufficient air speed had been reached in order to attain

Assam Trucking Company

airborne altitude. Pilots should have applied power constantly and smoothly as soon as the takeoff roll began.

Once the pilots became used to the idiosyncrasies of the aircraft, the accident rate dropped. When recommendations made to correct the problems of night takeoffs were implemented, the accident rate was reduced sharply. [14]

One quote associated with the C-46, "Put six handles on this airplane and you'll have the perfect flying coffin," was the response of one Hump pilot when asked to fly a plane he didn't trust. [16] Nicknamed "Dumbo," it was considered by pilots to be as dangerous as the Hump terrain and weather. With all its faults, the one definite advantage when the plane made it through to China, as it often did, was that its large cargo compartment hauled four tons of gasoline and/or other supplies. [17] Despite its problems, the C-46 carried over one half of the total troops moved by ATC. The Tenth Air Force also used the C-46 to transport troops. [18]

6.2.1 Technical Specifications:

The last of the "tail draggers," the C-46 was bigger than a B-17, with half the engines to carry approximately the same gross weight.

Engines: Two R-2800 Pratt & Whitney w/Curtiss four-bladed, steel electric props
HP: 2,000 (B-17—HP 3,600)
Dimensions:
  Wing span: 107 ft 2 in.
  Length: 76 ft. 4 in.
  Height: 21 ft 9 in.
  Weight:
  Max T/O: 48,000 lbs.
  Empty: 30,000 lbs.

Capacity:
- Crew: 4-5
- Troops: 40
- Stretchers: 30
- Cargo: 15,000 lbs.

Speed:
- Top: 240-245 mph
- Cruising: 160-170 mph
- Climb: 1.175 ft./min.
- Range: 3150 miles
- Ceiling: 25,000 feet

6.3 C-54, SKYMASTER

Among the other aircraft, C-54 Skymaster was a solid medium-range transport as an adaptation of the passenger-type DC-4. Like the C-47, it was built by Douglas Aircraft Company in Chicago, IL. It carried a payload of 9,000 lbs. at extreme range and 10,900 for a 2,400-mile trip.

The C-54B model was updated to remove fuel tanks in the cabin and install extra fuel tanks in the wings. The move increased cabin space and reduced the fire hazard; the passenger capacity was increased from 30 to 49, and litter capacity for air evacuation of the wounded. From 24 to 36 bucket seats added to the cabin increased the passenger carrying capacity from 30 to 49. For air evacuation, the litter capacity increased from 24 to 36. Bucket seats were replaced by canvas folding seats along each wall for a weight savings of 7 lbs. per seat.

The C-54C was unique: built and equipped especially for the use of President Roosevelt. The C-54D, was a modification with more powerful engines. It came into use in

Assam Trucking Company

August 1944. The C54-E, a luxurious passenger model and the C-54G, the corresponding cargo model, were not available until 1945. [19]

The Douglas C-54 found its Hump role at the southern end of the Himalayas, where the peaks were lower, but the mission distances were longer. Its 17-ton useful load usually consisted of 55-gallon drums of aviation gas. The C-54's, with four 1,350 hp Pratt & Whitney R-2000-7 radials, afforded a maximum range of 3,600 miles at 190 mph. Accordingly, the Skymaster flew the last of the Hump missions, carrying occupation troops to Shanghai after Japan surrendered. [20]

In August 1945, ATC had 839 C-54s of all models (A, B, C, D, E, and G) in service.

6.3.1 Technical Specification:

Crew:	4
Capacity:	50 troops
Length:	93 ft 10 in (28.6 m)
Wingspan:	117 ft 6 in (35.8 m)
Height:	27 ft 6 in (8.38 m)
Wing area:	1,460 ft² (136 m²)
Empty weight:	38,930 lb. (17,660 kg)
Loaded weight:	62,000 lb. (28,000 kg)
Max. T/O weight:	73,000 lb. (33,000 kg)
Powerplant:	4 × Pratt & Whitney R-2000-9 radial engines, 1,450 hp (1,080 kW) each

6.3.2 PERFORMANCE:

Speed:	
Maximum:	275 mph (239 kn, 442 km/h)
Cruise:	190 mph (165 kn, 310 km/h)
Range:	4,000 mi (6,400 km)

Service ceiling: 22,300 ft (6,800 m)
Wing loading: 42.5 lbs./ft² (207 kg/m²)

6.4 B-24 CONVERSIONS

6.4.1 C-87

The C-87 airplane was first produced by Consolidated Vultee at Ft. Worth, TX, in 1942 in a design competition for a fast, long-range transport. The adaption was called the "Liberator Express." Crews nicknamed the conversion the "Lumbering Lib." In addition to a flight crew of 5, it carried twenty-five passengers or an alternate cargo load.

6.4.1.1 Technical Specifications:

Engine: Pratt & Whitney R-1830-43 Twin Wasp radial engines
Power: 1,200hp
Crew: 5
Passengers: 25
Dimensions:
 Wing Span: 110ft
 Length: 66ft 4in
 Height: 18ft
Weight:
 Empty: 38,000 lbs.
 Gross: 64,000 lbs.
Speed:
 Max: 306 mph
 Cruising: 278 mph at 25,000 feet
 Climb rate: 20.9 min to 20,000 ft
Ceiling: 31,000 ft
Range: 2,900 miles

Guns: none
Payload: 12,000 lbs. cargo or 25 passengers
Engines: Four Pratt & Whitney R-1830-43 with exhaust driven turbo superchargers

6.4.2 C-109 TANKER CONVERSION

The B-24 Liberator bomber conversion was pressed into service as a cargo hauler (C-87) or a tanker (C-109 fuel hauler). These conversions were made possible by removing most of the armament and rigging of the bomb bay section to accommodate space for seats and cargo. Because of the problem of gasoline leakage, the C-109 was often called the "C-1-0-Boom," ... "in other words it tended to blow up before you could say C-109!" The first 15 C-87s arrived at Chabua, Assam, India on Christmas Day 1942. It was noted in the Unit History of Chabua that the planes arrived without instructions, Technical Orders (TOs—maintenance manuals), spare parts or tools. Along with the missing equipment, there were no trained maintenance personnel. [21]

6.4.2.1 Technical Specifications

Crew: 4
Span: 110 ft
Length: 66 ft 4 in
Height: 18 ft
Empty Weight: 35,500 lbs.
Gross Weight: 65,000 lbs.
Cruising Speed: 300 mph
Guns: none
Payload: 2,036 gallons of fuel
Fuel Capacity: 2,900 gallons

Mohanbari 1332nd Army Air Force Base Unit (AAFBU) reported much of the same problems: lack of spare parts, maintenance tools, flood lights and bulbs, drop cords, and generators. To avoid enemy action only daylight Hump flights were flown, leaving only the night hours for maintenance. Without the floodlights, maintenance was accomplished using flashlights. Adding to the problems was the lack of trained personnel. Cannibalism of other airplanes grounded for repairs provided what spare parts could be scavenged. Generators, electrical equipment, and hydraulic parts were in very short supply.

The C-109 was outfitted with 6 extra fuel tanks—one in the nose, two in the bomb bay and three in the rear fuselage—giving it a 2,400-gallon fuel capacity. The C-109 was not utilized to the same extent as the C-46 and C-54 for the movement of fuel. It was not as reliable as the other aircraft in theater. The modifications were made to 218 aircraft, with the majority serving in CBI.

A total of 276 Liberator Express transports and tankers were delivered during the two-year period.

ILLUSTRATIONS SECTION

Figure 1 Japanese- Controlled Areas, 1937-45

Figure 2 Proposed Land Route, Iran

Figure 3 Proposed Air Route

Assam Trucking Company

Figure 4 Air Transport Command Emblem

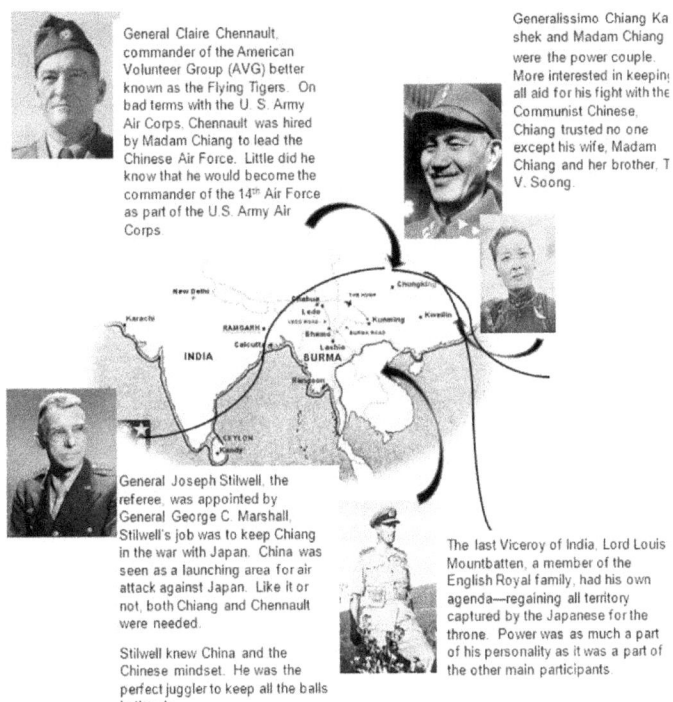

General Claire Chennault, commander of the American Volunteer Group (AVG) better known as the Flying Tigers. On bad terms with the U. S. Army Air Corps, Chennault was hired by Madam Chiang to lead the Chinese Air Force. Little did he know that he would become the commander of the 14th Air Force as part of the U.S. Army Air Corps.

Generalissimo Chiang Ka shek and Madam Chiang were the power couple. More interested in keeping all aid for his fight with the Communist Chinese, Chiang trusted no one except his wife, Madam Chiang and her brother, T V. Soong.

General Joseph Stilwell, the referee, was appointed by General George C. Marshall. Stilwell's job was to keep Chiang in the war with Japan. China was seen as a launching area for air attack against Japan. Like it or not, both Chiang and Chennault were needed.

Stilwell knew China and the Chinese mindset. He was the perfect juggler to keep all the balls in the air.

The last Viceroy of India, Lord Louis Mountbatten, a member of the English Royal family, had his own agenda—regaining all territory captured by the Japanese for the throne. Power was as much a part of his personality as it was a part of the other main participants.

Figure 5 Power Players

Figure 6 C-46 on Letdown/Approach in China

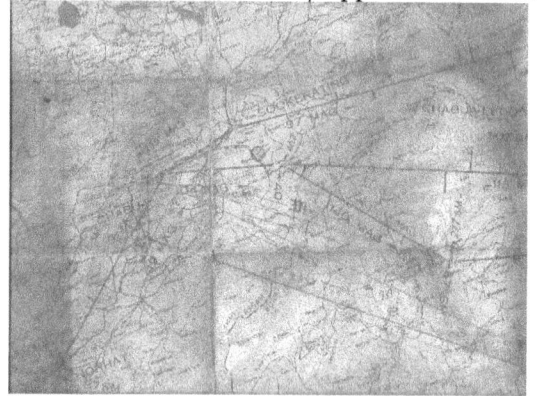

Figure 7 Assam Charts

Figure 8 Solo Flight Certificate

Assam Trucking Company

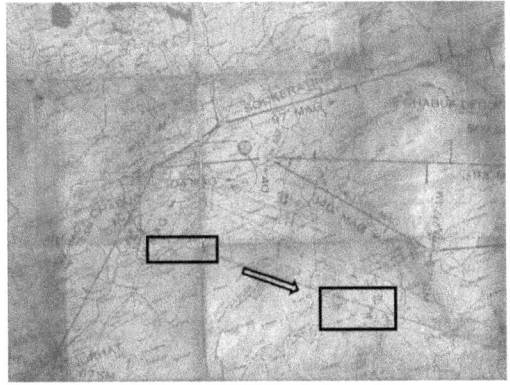

Figure 9 East Route Eastbound

Figure 10 British Tent and Basha

Figure 11 Butler Hangar

Figure 12 Miss Amari Loading Aviation Gas

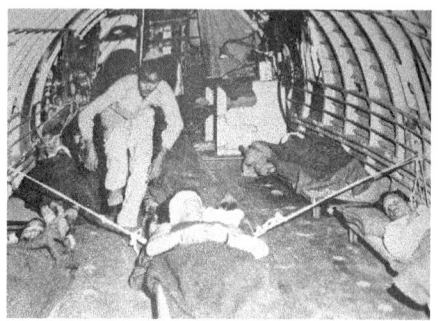

Figure 13 Med-Evac

Figure 14 F/O John Foster

Assam Trucking Company

Figure 15 Indian Bearers

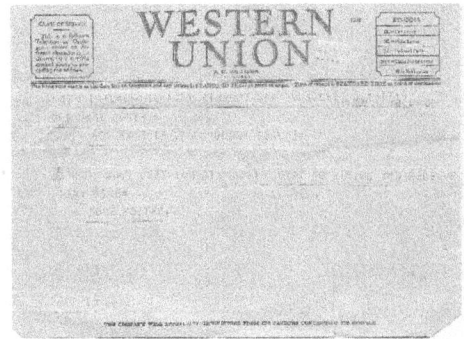

Figure 16 Home Soon

Chapter 7.0 — Maintenance

In 1943, maintenance was a challenge due to an overabundance of weather, usually wet (1945 set a record for rainfall in the Assam Valley—500 inches!), malarial mosquitos, Japanese bombing raids and lack of:
- Trained personnel
- Spare parts (at Misamari 13,000 spark plugs were used per month—depot sent refurbished spark plugs that often fouled)
- Maintenance/repair equipment and/or tools,
- Technical manuals and documentation.

Throughout 1943 and 1944, little changed in the maintenance arena. Because CBI was considered a backwater operation and was therefore low on priority lists, there was always a lack of personnel, parts, equipment, and tools. Local Indians were pressed into mechanic training to support maintenance at several base units:
- Misamari—15 native laborers assigned to the aircraft shops were expected to develop into competent shop helpers. Thirty more were to be trained in 100-hour inspection using PLM.

Assam Trucking Company

- Chabua—237 Indians were employed in aircraft maintenance. One half of the 237 underwent tool and elementary mechanics familiarization training.
- Jorhat—the mechanic shortage was met by assigning Kachin troops to the maintenance group, with very satisfactory results.
- Lalmanir Hat—Indian locals were trained in basic mechanics, freeing the maintenance personnel for systems maintenance and repair.

Wherever locals were hired, there was always the very real threat of Japanese sympathizers who could and would sabotage aircraft on the ground.

At Chabua in early 1944, a lack of organization in aircraft maintenance led Capt. Ralph Tilney to divide the unit into sections responsible for certain tasks—each with a specialty. In what proved to be a forerunner of Production Line Maintenance, crews were arranged to have 3 from the original crew assigned to one aircraft on the flight line, while the remaining members were assigned to a specific stage in the inspection process. A maintenance engine stand made of 3-inch steel tubing was built to allow a maintenance mechanic to get to any side of the engine without getting down on the ground. A lower platform built into the stand allowed men to reach the bottom of the engine. Special lights and drop cords were also installed. A large-enough slot in the stand was provided to allow the prop to be pulled through.[1]

Maintenance crews worked outside regardless of the weather or time of day. There were no hangars, only revetments. When a plane returned from China, the pilot would give the maintenance crew chief a list of discrepancies noted during the flight. Those discrepancies and any regularly scheduled inspections, e.g., 100-hour inspection, etc., would be handled by a maintenance crew assigned to that aircraft. Other maintenance crews would stand around awaiting assignment to a plane. The crew chief and his maintenance crew were concerned with only one plane. Because tools and equipment were also in short supply, a specialized tool was often not in the area where it was needed, but in another revetment. There was no organization. At best, maintenance was hit and miss. [2]

If a pilot arrived at the plane ready to run a preflight check and the crew chief was available, the pilot would ask if the crew chief was going on the flight. If the answer was affirmative, the preflight was dispensed with. If the answer was negative, a full preflight would be completed. The mindset was that if the crew chief intended to go on the flight, all must be well. [3]

Gordon Smith spent a year as crew chief on a C-46 often in the copilot seat as the copilot and radio operator readied for bailout. Generally, he noted the problem was with the engines. If the plane was empty on the westbound leg and they lost an engine, there was no problem. He could get the plane back to base. His feeling was that a lot of planes were lost on judgement calls, not because of a problem with the plane, but mostly due to inexperience of the flight crew and the maintenance crew. Smith preferred to fly than to bail out and walk out. If the crew chief felt the plane was safe to take out but the pilot didn't have the same level of confidence, there would be a conversation between the Chief and the tower. Sometimes the plane took off with the assigned crew or another crew would be called out. The pilot who rejected the aircraft would be "chewed royally."[15]

Others like M/Sgt Rachel, a line chief on the C-46, said of the plane, "Too much airplane and too many bugs." Rachel started classes, teaching maintenance crews what he had learned. Because of his ability to understand and pass on his working knowledge of the C-46, and for the resourcefulness in making engineering equipment, Rachel was recommended for the Distinguished Service Medal.[7] The capabilities and operating characteristics of the C-46 were unknown to maintenance crews. Most had trained on the C-47, not the C-46.

7.1 PRODUCTION LINE MAINTENANCE (PLM)

Late in 1944 a new concept was being researched and considered for use in CBI that would change maintenance operations—Production Line Maintenance (PLM). Based on the automotive industry's approach of the assembly line, PLM was considered to be the answer to the maintenance problems encountered in theater. When PLM was first introduced at Jorhat AAFBU 1330 in February 1945, the aircraft slated for the first-ever approach anywhere overseas for the ATC was a C-87. On February 16, 1945, Brigadier General William Tunner, the force behind the establishment of PLM, inspected the facility and praised the efforts being made. Tunner felt that the approach would aid in better maintenance with shorter turnaround times, resulting in more trips over the Hump. More trips conducted with fewer problems led to increases in tonnage delivered to China with less loss of planes and lives.

At Misamari, Administrative Officer Lt. Henry C. Bush reported that PLM begun in February, with the main savings in time and effort. "Heretofore, specialists romped from one hardstandings to another working on ships. Now ships" would be brought to the maintenance crews at their specific revetments.

Specialized crews were placed to conduct specific tasks. For example, the first crew removed the cowling and prepared the engines for maintenance; then next crew had specific functions to complete before passing the plane on to the next maintenance station. With a specialized crew at each station concentrating on one maintenance function, crews became more proficient at their jobs and were able to perform their tasks more quickly and with fewer errors. No longer did they have to search for a specific tool. Tools needed for particular functions were placed at the applicable station. [4]

With all new systems and approaches, PLM was not without its problems and detractors. The new system was supposed to decrease the time an aircraft remained in maintenance and to increase the number of aircraft in operation per day. At this time several issues were working against the concept:

- Sudden introduction of the maintenance approach
- New aircraft being received and integrated into the operation
- New, untrained personnel introduced into the system
- Continued high rate of out-of-base engine changes
- Unpopularity of the new approach to maintenance by both maintenance personnel and pilots.

To this last point, maintenance crews felt a loss of camaraderie, having previously operated as a specialized crew of 15 men (including flight engineers). Under PLM they were being assigned one routine job per man. Each man had a specific duty "which he performed continuously with resulting monotony and lack of interest in the job." [4] The monotony was relieved by rotating the jobs, resulting in a more efficient inspection staff. [5] Additionally, the rate of

Assam Trucking Company

aircraft losses increased toward the end of February, producing a loss of confidence in the program.

Gordon Leonard, a check pilot stationed at Jorhat 1944-45, remembered when PLM was instituted. The demand for maximum employment of aircraft required a different maintenance plan, and the term was Production Line Maintenance (PLM). Whenever the aircraft returned from a flight, all of the discrepancies noted during the flight were processed, with maintenance trying to correct all the problems. In addition, on a periodic basis, various scheduled maintenance tasks were carried out. The new approach made the aircraft essentially available for a flight most of the time. Maintenance was conducted in all kinds of weather, at night, outside, with mosquitos chewing on the maintenance recruits. Despite these conditions, records were established of aircraft utilization per month that approached 300 hours, which was considered impossible in a stateside environment at that time.[6]

The Unit Report from Chabua in March 1945 recommended that Indians not be used as mechanics because they did not understand English, did not follow orders, and did not show up for work. It was felt that Whites and Blacks assigned to general duties should replace the Indian workers. "Blacks should be trained as mechanics, since they showed more interest. Their work has been excellent with supervision in PLM and engine change."[7]

The lack of trained personnel at Misamari in 1945 was alleviated in part when 15 native laborers were assigned to the aircraft shops. They were expected to develop into competent shop helpers; however, the trial was a failure. Thirty more were tried in the 100-hour inspection routine of PLM. Like the previous attempt, the Indian laborers didn't work out.[8]

7.2 PLM Stages

Col. Robert R. White, Chief of Aircraft Maintenance in ICD, Col. William S. Barksdale, Jorhat CO, and Capt. William P. Dunn, Base Director of Aircraft Maintenance, worked together to develop the new PLM system introduced at Jorhat. The five stages developed were:

- Stage 1—Engines de-cowled, cleaned and inspected; plane washed or sprayed and carburetor, props, engine mounts, conduits, hydraulic and vacuum systems inspected
- Stage 2—Thorough testing and renovation of radio, electrical and oxygen systems
- Stage 3—Inspection of power-plant accessories and safety devices; entire aircraft rechecked to ensure corrective actions had been completed; sheet metal repairs made
- Stage 4—Aircraft jacked and landing gear, including retraction of the wheels, inspected; final inspection of engines.
- Stage 5—Engines re-cowled and maintenance engine run with instrument checks; final review of all systems

Because the stages were based on aircraft systems, i.e., engines, hydraulics, electrical, etc., systems training was emphasized to support the new approach.

Construction of Butler hangars began in coordination with the PLM system. All the work was completed in a series of Butler combat hangars erected to provide the coverage needed to complete maintenance activities out of the elements on a 24/7 schedule. The Butler combat hangars (canvas-covered frames) provided

- Protection from:

- Monsoon rains
- Excessive heat
• Provided adequate lighting while limiting glare.

Under the previous system, 100-hour inspections could only be accomplished during daylight hours outside, and took an average of 36-hours to complete over a 3 to 4-day period. Under PLM, with the use of work stations in Butler combat hangars, the 100-hour checks were accomplished in a more efficient manner in fewer days. The goal was to reduce the check from 36 to 15 hours. One hundred-hour checks were not accomplished in China, adding to the backlog of inspections to be done at the Indian base units and a decrease in delivered tonnage. [9]

In April 1945 the Butler hangars used in PLM were completed at Misamari, AAFBU 1328. Two airplanes could be housed in one hangar for 100-hour inspections. The hangars were arranged to allow the taxiway to extend through the building permitting planes to pass through. The aircraft could leave either end of the hangar without causing congestion. Use of the combat hangars released a number of tugs and men from moving aircraft. Butler hangars were received at Mohanbari in May 1945. PLM had been started at Mohanbari March 21, 1945. Until Butler hangars were in place, only nose-hangar facilities were available. [10]

Despite PLM, part shortages continued. One plane was grounded from 4/1 to 4/8 1945 waiting for landing gear parts. Engines were in short supply as well.

The first month PLM was in action, Jorhat reported there were 93 white and 36 Black enlisted men working in PLM. A breakdown of the personnel follows:

DUTY	WHITE	BLACK
Line Chief	1	
Flight Chief	3	
Power Plant Crews	65	
Hydraulic	6	
Heaters	2	
Shrouding	2	1
Plugs	3	4
Fuel-Electric-Cockpit	2	
General	1	3
Aero Repair	1	
Dispersal Chiefs	3	
Cable	2	2
Wash Rack		15
Retraction		3
Greasing		8
Stock Chaser	2	
TOTAL	93	36

Table 7-1, PLM Personnel [11]

Despite the high expectations for the success of PLM, aircraft maintenance efficiency at Jorhat was difficult to analyze for two reasons:
- Introduction of the totally new PLM system

Assam Trucking Company

- Arrival of 171 new enlisted maintenance men, most untrained in maintenance of C-87- and C-109-type aircraft

PLM was initiated to reduce inspection time per aircraft from 36 hours to 20 hours (exclusive of major overhaul, e.g., engine changes, etc.), increasing the number of inspections. However, a look at the statistics for February when matched to the figures from January, revealed that the time to keep aircraft in flying condition increased to 11.9 hours from 9.72 per plane per day.

The disruption in maintenance activities caused by the changes in maintenance procedures and the new personnel were not the only differences. The number of aircraft assigned daily during February increased to 41.5 as compared to 37.9 in January and the daily operation aircraft went from 31.5 in January to 33.6 in February. There were also a number of out-of-base engine changes. At the end of February, it was too early to determine if the new program was a success.

As with all new programs and procedures, there were naysayers among the maintenance personnel and pilots. The absence of crew spirit was cited as the problem among maintenance personnel with PLM. Now specialized crews of 15 men each, including flight engineers, performed the same routine job assigned, which they claimed resulted in monotony and lack of interest. PLM crews were placed on round-the-clock rotations with 24-hour availability. [12] There seemed to be no challenges.

The increase in aircraft accidents toward the end of February after the introduction of PLM produced a lack of confidence among the pilots in the maintenance program.

Despite the unpopularity of the program and the increase in time for inspections, one bright spot existed. No aircraft had been grounded during the month because of a lack of parts, compared to five planes being grounded during the month of January. The perfect month with no APOCs (Aircraft part -Out of Commission) resulted from close coordination with the 329th Service Group and other bases.

Adding to the maintenance tasks was the damage caused by vultures. A known flying hazard, the large birds often came into contact with aircraft on approach or takeoff. A sizable bird with no feathers on their long necks, vultures feasted on carrion. It was not unusual for a vulture to contact an aircraft on the leading edge of a wing, an engine or even the cockpit, killing the pilot. With a wingspan of 8 to 8.5 feet and weighing as much as 20 pounds, vultures could and did do a great deal of damage. In one instance, an aircraft aborted after an encounter with a vulture. When the pilot taxied in, all that could be seen of the vulture was its head with its tongue sticking out and one clawed foot protruding from the leading edge of the wing of the airplane. A rope secured to a "tug" used to tow aircraft, was used to pull the bird out of the wing. The sheet metal repair was completed with an aluminum patch. [13] One pilot reported seeing 21 feet of a wing taken out by a vulture.

On November 9, 1944 an inbound C-109 from the States to Kurmitola, experienced a "buzzard" hit on a slow descent for landing. The buzzard, having not been trained on the rate of closure, came through the left side of the windshield, hitting the copilot and the navigator. They both ended up on the flight deck, wounded and bloody, the coplot having sustained 30 lacerations and the navigator 20. The radio was knocked out, requiring the pilot to signal the tower's attention to get a green light to land. The bird's wingspread was measured at 14 feet. The accident report stated that the pilot should have zigged when he zagged. [14]

7.3 ENGINES

A new type of engine for the C-46 was received at the Depot at Agra in 1943 without notice. The differences between the old and new engine types caused problems until Technical Orders were received. The common problem of new aircraft deliveries without T.O.s or parts peculiar to the aircraft was accompanied with a lag time between the delivery and the support manuals and parts. For any of the operations, lag time, if not given enough lead time, could cause critical shortages, endangering mission success. Command felt it would be beneficial to ship aircraft with T.O.s and enough parts to cover contingencies until a supply could be established, e.g., green wing-tip lamps for the C-46. The C-46 parts shortage was an ongoing challenge. Of particular note were shortages of prop governors, prop governor gaskets, hydraulic pressure regulators and spark plugs, integral parts to the engines. Tiedown equipment shortage was critical.

In June 1945 there were an unusual number of emergency engine changes in China and home bases. The percentage rate of failures was noted in overhauled engines. The life of an engine overall had fallen to 330 hours, with no discrepancies recorded in maintenance procedures regarding care and installation. [15] The environment of India and China with the humidity, dust, changes of weather, the icing, and other flying conditions, all contributed to the wear on and increased failure rates of aircraft engines across the board.

Chapter 8.0 — Logistics

From the opening of the CBI theater through the end of 1943, there was a policy of no clear night flying, which was dependent on runway lights. Three fields, Chabua, Misamari and Mohanbari, received marker lights and spare parts for generators and flight instruments in September. At Mohanbari on the 1st of November, the C-47s assigned moved to the western sector of the ICW, leaving only C-46s in place. Round-the-clock operations were started.

As with any other theater in World War II, there were always snags that affected the mission. More so than the other theaters, CBI was totally dependent on airlift for all mission supplies. Bad weather caused planes to be grounded or lost over the Hump resulting in delays or loss of tonnage. The loss of a plane over the Hump not only wasted the cargo and aircraft, but possibly a crew as well. If it wasn't the weather, it was the Japanese shooting planes down or damaging them on the ground during air raids. Any loss was a step backward for the war effort against the Japanese.

At China terminals, Japanese air attacks were a daily reality. More than one transport crew spent the evening under attack or being waved off from a field with Japanese aircraft aloft. If a plane was waved off, their choices of alternate bases were limited. Other bases may be under attack as well, or were out of range because of reduced fuel quantities on board the aircraft.

Despite the weather, Mohanbari's tonnage in January 1945, was 7,718.184 tons completed in 1,793 trips. Mohanbari carried the most of any of the basses—19.2% of cargo carried. Sookerating was second with 6,653.631 tons.

The receiving bases of the tonnage in China were:
- Yunnanyi—267 trips (2140 tons)
- Chanyi—49 trips (392.135 tons)
- Chengkung—56 trips (448.84 tons)
- Yanghou—27 trips (217.35 tons)
- Luliang—72 trips (577.08 tons)
- Kunming—731 trips (5858.965 tons)
- Paoshan—258 trips (2067.87 tons)
- Tsuyung—95 trips (761.42 tons)
- Tenchung—201 trips (1611.015 tons)
- Chaotung—34 trips (272.51 tons)
- Mangshih—3 trips (24.245 tons)[3]

8.1 SAFETY PROGRAM

Under General Tunner's command, a Flying Safety Program was instituted. The seven points of the program were:
- Investigate fully the training of incoming pilots
- Check the weather as a means of combating existing conditions, i.e., icing, turbulence, computing wind velocity in good weather and bad
- Communications—use of radio compass, radio limitations, when and when not to use "May Day"

- Pilot Discipline and Airport Discipline—Checklists, pre-flight, and post flight
- Briefing and debriefing—competency of crew, thorough preparation of pilot for existing conditions en route. Debriefing—Competency, problem areas, corrections and training needed, best weather reports
- Maintenance
- Airport Facilities—upkeep important to safety

It was General Tunner's feeling that health checks should be made for each pilot before flying, and their diets regulated. "Eating gas-producing foods even hours before take-off would result in debilitating agony at high altitude." A crewmember not in good health could cause an accident just trying to get their points in for rotation home. One man could cause the loss of a plane and as many as 45-50 lives if the cargo was a group of soldiers. Pilots were monitored to ensure the safety of a flight. Each accident was to be thoroughly investigated. Maintenance procedures done on the plane prior to flight, weather conditions, and the pilot's activities scrutinized on and off duty. Pilot error was a primary cause of crashes, with the most susceptible: new, inexperienced pilots. Those over thirty with 2,000 hours or more of flying time were less susceptible. [4] To this day, the first step taken when investigating any incident is to check the training records of the crew, especially the pilot.

8.2 ATC UNITS

The lifeline established from Karachi (now part of Pakistan) to Dinjan (the CNAC base) and on to China, was activated in early 1942 and served as the western ATC hub of the India-China Air Route. Airplanes, supplies and personnel passed through the base from March 1942 until it was deactivated in late 1945.

Beginning in January 1945, passenger flights started. Scheduled Valley Flight 51, a C-47A with airline seats, left Chabua in early morning and flew to Misamari. Flight 52 took off from Lalmanir Hat and arrived in Misamari in late afternoon. A freight schedule for the Assam Wing operated on a similar schedule. [5]

8.3 OTHER UNITS

ATC was not the only group vying for supplies. Combat Cargo groups flew C-47s with camouflage paint as their only protection. They were responsible for supplying the British 14th Army land operations in Burma. Their mission was to keep the Japanese from moving into India at Imphal. Using the airbase at Imphal, pilots who had no flying regulations and rarely filed flight plans flew only 25 minutes from the base to the drop zone. They flew just 50 feet above the treetops to keep from casting a shadow. Any higher would alert the enemy to their presence. In 1942, they had 100 DC-3s and 300 pilots. [6] Combat Cargo carried out airborne resupply and evacuation missions of wounded, and used gliders for assault missions. Commando units would parachute at low altitude behind enemy lines, perform their mission, then either walk out to friendly territory or to a small number of C-47s which had clandestinely landed at a rough airstrip to pick them up.

8.4 CARGOS

Cargos fell into two categories—wet and dry. Wet cargoes included aviation gas; everything else dry, primarily munitions. A team of two enlisted men (usually Blacks) served as supervisors, plus a number of locals loaded the planes. Indian Pioneer troops typically wore steel-cleated, oversized shoes, which were considered a fire hazard when loading drums of high-octane gasoline. They were used on all dry cargo loads. [7]

8.4.1 AVIATION GAS

The main cargo hauled over the Hump was aviation gas in 55-gallon drums or on C-109s in leaky fuel tanks that had been installed. It was not unusual to see an elephant being used to load the drums of gas into a plane.

In 1945, gasoline and oil accounted for 60% of all net tonnage eastbound. Ordinance amounted to 15%. Seven 55-gallon drums were loaded forward of the center of gravity (CG) and seven drums loaded aft of CG. Other cargo was loaded between. The balance included placement of passengers and supplies for Air Corps technical, PX, and Quartermaster. Westbound aircraft had smaller loads; by 1945 they flew empty. [8]

8.4.2 VEHICLES

Vehicles, which were in short supply, had spare part issues as did the planes, and were not useable in monsoon weather from May to October. The unpaved roads were badly constructed, and mostly consisted of mud due to the average rainfall of 200 inches/year.

At times when trying to load a jeep or larger vehicle into a C-46, it would be cut in half or disassembled, loaded in the plane and reassembled in China.

Assam Trucking Company

8.4.3 PX SUPPLIES

At Chabua the PX opened once every two months with restricted (severely limited) supplies of razor blades, soap, and toothpaste. In late 1943, candy bars, beer, cigarette lighters and hair tonic became available.[9] At Mohanbari, PX supplies, never abundant, hit still lower levels when toilet necessities were unattainable.[10]

The word "limited" does not come close to describing the availability of everything needed, from clothing through the Quartermaster Corps, paper, office equipment and Post Exchange (PX) items to food and tents. Proper clothing for flight crews ranged from slow-to-be-delivered to non-existent. In December 1944, flight uniforms arrived. Before December 1944, the aircrews would wear whatever they could find. In the winter they would wear two or three layers of clothing—even blankets.[11]

Equipment sent from bases already in operation to others for start-up were slowly replaced if at all. If the equipment was returned, it was generally worn beyond use. Oxygen, necessary to fly at the altitudes to clear the lower spine of the Himalayas, was not available. O_2 masks were in very short supply. Oxygen had to be flown to the Assam bases from Calcutta. An oxygen generator plant was built later on in operations at Chabua. The plant exploded on more than one occasion, causing flights to be grounded or made without oxygen on board.[12]

8.5 REQUISITIONS

Capt. C. F. McLaren, Jr, Wing Supply Officer, reported in an interview that a large number of requisitions from China were missing, causing a shortage of supplies in China. The Air

Corp Supply attempted to trace back requisitions and expedite the flow of the supplies to China. Many of the requisitions were never received by the Depot at Chabua. During the investigation, no evidence was found to explain how the requisitions had been lost. The process in use was for China to send a requisition to the Depot at Chabua, which would then send the requisition on to Calcutta (Dum Dum). It was decided to cut out the middle man and send the requisitions straight to Dum Dum. This change was successfully implemented between March and April, 1945. [13] Removing the middle man at Chabua, eliminated the problem of missing requisitions.

8.6 CARGOES OUT OF CHINA

Traffic out of China included troop transportation for combat training in India. Strategic materials such as were considered essential to the war effort:
- Wolframite ore, a form of tungsten
- Tin
- Hog bristles
- Mercury
- Silk 14

Part of the reason the listed items were considered strategic was to keep them away from the advancing Japanese. Wolframite was refined in China and sacked, like flour, in five-pound bags. Tungsten was used in the U.S. to make armor-plate steel.[15]

8.7 PLANES

The first 15 C-87s arrived at Chabua, Assam, India, on Christmas Day 1942. It was noted in the Unit History of

Chabua that the planes arrived without instructions, Technical Orders (TOs—maintenance manuals), spare parts or tools. In addition, none of the mechanics available had any training on C-87 maintenance. Utilization of the C-87s was extremely limited.

Mohanbari 1332nd Army Air Force Base Unit (AAFBU) reported many of the same problems; a lack of spare parts, maintenance tools, flood lights and bulbs, drop cords, and generators.

Lalmanir Hat received its first C-46 on 10 January 1945, with another 17 by the end of that month. With the new aircraft came a tremendous turnover in personnel due to the change from C-47s. Overall there was an increase of 151 men—79 Officers and 72 enlisted. By the end of February 1945, even with the increase, the base was understaffed by 103 enlisted and 7 officers.

The mission of the base was to carry out flights from Barrackapore to the Assam bases in order to provide planes and crews. At the end of February 1945, the mission expanded to include hauling pipe to Burma from Deragon. In February of 1945 there were too many copilots for the number of pilots. Flights were delayed at Barrackapore due to excessive work load and lack of first pilots. The average waiting time was 6 hours and 45 minutes per trip. Fewer trips meant that less copilots were utilized. The flight check system was put into play to advance copilots to first pilot and thus balance personnel utilization.

8.8 PLANE PARTS

When the first C-46 arrived, the same problems as with the C-87 were evident. In addition, the new plane had not been completely service tested in the States. By the middle of

1943, the parts shortage was critical. The emphasis had been placed on planes, not on the needed parts. [20] In CBI, "there was no overall Allied air command which had authority to formulate air strategy. The war was being fought on the concept of cooperation instead of command, which did not work well, breeding endless problems and differences. Concurrence was needed among the RAF, Tenth Air Force and Fourteenth Air Force commanders for airfield development, disposition of units, selection of target systems, allocation of aircraft and air corps supplies, allocation of personnel, and related problems." There were too many cooks in the kitchen.

In ATC the testing of the C-46 had been done in relatively good weather and not in mountainous terrain. Because of the different environment and altitude requirements problems became evident quite early with the C-46, a bigger plane, causing some pilots to refer to it as a "flying coffin." Delivered late in the year, the plane hadn't even been winterized.

As aircraft had mechanical problems, some flights were grounded until parts for repair could be found. The problem with being at the end of the supply line was magnified. Some in higher levels didn't see the necessity to supply this backwater theater.

There was a great deal of competition over the supplies that arrived at the Assam Valley bases. A first come-first served mindset prevailed, and the fight was on. "The formation of headquarters, AAF, India-Burma Sector in July 1943 was just the first step toward solving the problems."

8.9 SHORTAGES

Shortages were part of the CBI Theater from the beginning. At Misamari, a shortage of construction materials was evident. The lack of nails, lumber and electrical wiring were held up 75% of all jobs of the Base Maintenance Department. The supply of bamboo from British sources also dried up, resulting in the stoppage of repair to existing buildings and new crew quarters construction.

The Base Snack Bar at Misamari opened on 7 April 1945, with a soda fountain, ice cream storage and hardening cabinet, coffee maker and drink mixer. It did double duty until they ran out of some items.

8.10 CHINESE TROOPS

The movement of Chinese soldiers unaccustomed to the concept of flight was always a challenge. One of the lessons learned was to take all weapons (knives, grenades, etc.) from the Chinese soldiers and secure the weapons in the belly compartment of the plane.

In one situation, "one planeload of Chinese decided that the pilot and copilot had given them a rough ride on purpose. After landing, as the crew stepped off the plane, they were surround by a group of angry Chinses with bayonets affixed to their rifles. They were about to kill the crew when an English-speaking Chinese officer diffused the situation.

"Other troop-carrying planes would arrive at its destination with one or two men short. It seems the Chinese had developed a new game. They considered it a big joke to open the cargo door, entice somebody to it, point to something interesting below, and then push out the unsuspecting soldier.

"Another plane appeared to be shedding Chinese as it sped down the runway. At the end of the runway, the craft went airborne and was gaining altitude when it suddenly flew straight up then flipped over on its back. All on board died.

It was obvious what had happened. The Chinese panicked and ran to the tail of the plane to get out, causing the plane into a nose-up attitude. Weight and balance being what they are, the tail was over-weighted flipping the plane on its back. Gravity took over from there.

"If locked into the plane, unable to get out, the Chinese would try to get to the crew. It became part of procedure to ensure the cockpit door was securely locked. Once airborne, to sooth the fearful passengers, the pilot would climb to altitude while the crew donned their oxygen masks. Once at altitude, the Chinese, suffering from anoxia, peaceably went to sleep.

"It gets icy cold at the altitudes required to clear the mountains and the Chinese were uncomfortably cold. Some passengers were known to set fires in the middle of the plane.

"For other crews the stench was worse than the dangers of transporting Chinese recruits. They weren't the cleanest people possible, and the planes had no sanitary facilities on board. As a result, the passengers would urinate and defecate where they chose. Airsickness then kicked in, with the Chinese spewing half-digested rice all over the plane and each other for five hours."

Lessons learned resulted in:
- Disarming incoming passengers
- Sitting arrangement on the floor with tie-down rings to use during takeoff
- Explaining any bouncing of the plane was NOT the pilot's fault
- Building of fires not permitted
- Providing a 55-gallon drum for bodily functions

While some aspects were unpleasant, the Chinese transported made the Hump mission easier in the end."[21]

8.11 EFFECTS OF LOGISITICAL PLANNING

The detailed coordination of complex processes and procedures was key to the function of all aspects of Hump operations which supported the airlift. Directly impacting the cargoes carried and the flights to available units in the Command's planes was the support of the:
- Movement of war matériel, troops, and supplies among the various bases and units
- Search and rescue efforts for downed aircrews
- Medical evacuation of injured troops and POWs
- Training and safety programs

The effects of the logistical planning made the difference in the growth and efficiency of the airlift mission.

Chapter 9.0 — Operations

As ATC units became functional, various movements of Chinese troops to the Indian training centers took place along with the opening and supply of Chinese interior airfields. The front in Burma was supported with airdrop of much-needed provisions. C-47s and C-46s of General Stratemeyer's Eastern Air Command (EAC) made daily runs, dropping food, ammunition, weapons, fuel, and medical supplies to frontline positions in Burma. Allied forces, roughly equivalent to three armies, fought for the most part with airborne supplies. [1]

On March 1, 1944, the American Army Railway Operating battalions took over running the Bengal-Assam Railway from the Indian government. The drive was on despite the Japanese offensive to cut or slow supply. Tonnage increased forty percent. [2]

As part of the effort to recapture Myitkyina, the "river, rail, and highway communication center" of North Burma and key to the building of the Ledo Road the oil pipeline to China and a refueling stop, the ATC played a vital role in airlifting the Fifth Indian Division, its equipment, men and pack animals to the front. A move that would have taken months now took only hours. [3]

Assam Trucking Company

As the number of available aircraft and manpower increased and Myitkyina was recaptured in August 1944, Hump operations were at last stepped up to meet the needs in China. Starting in October of 1944, Hump tonnage jumped from 18,000 to 24,000 net tons, an increase of 5-10,000 tons per month. It was estimated that to keep the Chinese in the war against Japan, 70 liberty ships would have been needed to deliver the tonnage necessary for the mission. One problem with that approach was the Japanese occupation of the major Chinese ports. These increases finally made it possible to stockpile some war matériel in order to launch a "back door offensive" against the Japanese.

In July 1943, only 2,916 net tons were airlifted to China. July 1944 saw that figure rise to 18,975 net tons. Then in July 1945, all existing records were broken, with 71,042 net tons of supplies transported over the Hump by the ATC and other units pressed into airlift service under control of ATC. [4]

As the hub of ATC in India, Chabua was to the Hump flights as Kunming was to intra -China movement of military cargoes. In October 1944, the impetus began to provide a solution to the problem of quickly transporting supplies and personnel across the vast areas of western China. Roads were few there and motor fuel and vehicles almost non-existent. Transportation within China was also the most dangerous element of the airlift, with few navigational aids, unreliable maps and the ever-present icing problems plus marauding Japanese fighters. Many of the peaks in China rose over 10,000 ft.

The Japanese had launched an attack from Manchuria in December 1944 toward the vital forces of the 14th Air Force bases in northern China. Defense of Chungking and Yunnan Province required ATC participation. First complete facilities, troops and equipment to be moved were identified.

Then a temporary field was established. When the field was ready, thousands of troops were flown from northern China to Yunnan. The 10-hour round trips were made using instruments in bad weather. [5].

By 1945 gasoline and oil accounted for nearly 60% of all net tonnage, while ordinance amounted to 15%, with the balance in passengers and supplies for Air Corps technical, Post Exchange (PX) and Quartermaster. Smaller loads were to be found on westbound flights to India. Traffic out of China, when encountered, consisted of troops, many of which were Chinese en route to India for training, wounded or ailing personnel.

Table 9-1, Tonnage Delivered by India-China Division—Air Transport Command

1944	To China	Intra-India
October	24,715	12,224
November	34,914	15,553
December	31,935	16,249

1945	To China	Intra-India
January	44,099	17,112
February	40,677	17,118
March	46,45	19,424
April	44,254	19,569
May	46,394	15,015
June	55,387	14,269

Myitkyina in northeastern Burma had been the key to this success. The recapture of this vital jungle outpost made it possible for the transports to carry more fuel and supplies to China. Myitkyina became a refueling stop for westbound traffic. The Ledo Road, another key supply route, was finally completed, as was the petrol pipeline. Although the Ledo Road went all the way to Kunming, China, the pipeline stopped at Myitkyina. [6]

9.1 CREW SCHEDULING

Col. Robert H. Baker, Commanding Officer of ICDATC, and Major Charles O. Galbraith, Director of Operations, developed a Standard Operating Procedure (SOP) for the turnaround crew flight schedules based on thirty-six-hour rotations. A round trip over the Hump with unloading time in China equaled approximately 10-12 hours.

Crew scheduling was based on "first in, first out," which was determined by landing time in accordance with Daily Operations Training, except for route checks and en route training. The composition of the schedule was made from a card file divided into the following categories:

- 1st Pilots available
- Copilots
- Administrative pilots and copilots
- Extra duty
- Pilots on trip
- Copilots on trip
- DNIF (Chabua)
- On Leave
- Special Missions
- Miscellaneous (missing, etc.)

Upon return from China, each crewmember's card was placed at the back of the card file, setting up a simple rotation. Crews that "sweat" (sat on the aircraft and waited for clearance for takeoff) a ship (Chabua), or had not departed in four hours, were released and another crew called. The first crew would be available eight hours after release from operations. Other crew personnel assignments were made from a list of flight radio operators in order of assignment. The list was submitted by the Communication Section to the Operations Officer, who then made the schedule.

When called, the crew was to report to the Orderly Room within 30 minutes and sign in. Fifteen minutes after receiving the crew call slip, the Transportation Dispatcher at Operations sent a driver/vehicle with slip to the Area Orderly (AO) room to pick up the crews. Slips were checked at the AO room, at which point the driver returned the crew and slips to the flight operations.

At Ops, their work was just beginning. The Pilot received the latest weather report from the forecaster, a briefing on Priorities and Traffic, then signed two copies of the load computation and manifest. The whole crew received a briefing, during which they were given a navigation kit, the latest NOTAMS and radio briefing materials. Briefing clearance materials issued included:

- Handbook
- Map set
- Radio Facility chart
- MB-4 Aerial Navigation Computer
- Weems Plotter
- Weather Code
- Verification Code
- Identification colors
- ICD-ATC Regulation No. 63-1
- Code Transfer to Tower
- First aid kit
- Duplicate copy of briefing clearance
- Plane Load Computation1-1289 AF ATC Form 8-31-43

- USAAF Cargo & Mail 24 Aug 42 Manifest, AAF Form No. 968
- Draw out money belts

During operational approval, the pilot signed a clearance and presented it to the Ops Officer for signature. He then signed the Form F (Weight and Balance) and reported to the Transportation Dispatcher with his request for transportation to the aircraft. Crews were to be at the plane 30 minutes prior to departure. To keep running of engines to a minimum and ensure an orderly movement of ships from revetments to the runway, crews were required to call the tower for permission to start engines.

9.2 INCOMING DEBRIEF

Upon return from China the crew reported to operations for a debrief.

Pilot turned in to the Operations Desk the following:
- Money belts
- First aid kit and briefing kits
- Original copy of the load computation form and manifest
- Flight clearance
- Form 1 with the time and any noted problems with the plane that required maintenance action
- Pilot Trip Summary Report

The radio operator turned in and signed off for the IFF.

The crew submitted an in-flight weather report, along with any information on new crash sites, enemy actions or anomalies. The pilot submitted a request and reported to the Transportation Dispatcher to order transportation back to the housing area. The pilot's card was moved to the back of the box for rotation.

The assignment to ATC was such that most of the men had few friends. In ATC there were no hard crew assignments, much as in today's Air Mobility Command and any of the major air carriers. A pilot would show up in base operations not knowing who else would be in the crew. There were common eating facilities; the huts and tents were only a few feet apart. They knew their basha/tent mates, but few beyond those quarters. They flew with one purpose—to amass the hours needed for rotation back to the States.

For the ATC crews, the primary need was for 1st pilots. A 1st pilot was given a minimum of 1 route check, which in the schedule was labeled "night route check" so that the landing in China was after dark and the crews were called for show up time. Recently upgraded to 1st pilot, the former copilot had one run before being placed on the schedule.

In March 1945, Approach Control Units (ACU) were established at Chabua, Mohanbari, Sookerating, and Myitkyina in preparation for the coming monsoon season. A control area was set up covering a radius of 25 miles from each field. The concept of the ACU was to expedite arrivals and departures of aircraft operating under conditions of reduced visibility, and to relieve control towers of the responsibility for instrument letdowns. It worked well once the pilots got the hang of it. When ACU was not operational, it caused trouble with letdowns over the beacon.

Icing conditions continued, necessitating special clearances. Weather and inexperienced pilot personnel were the major cause of accidents during the month of March. Pilot error in weather was given as the cause of eight accidents, with 19 aircrew fatalities, 17 missing and 52 Chinese soldiers killed. Great concern was expressed over pilot error problems in conjunction with a high attrition rate. Structural failure was also a concern. Maintenance, "supervisory laxity," negligence and casualness were mentioned as possible sources of trouble. In March there was also a diversion of Hump aircraft for special missions. Four flights daily were noted as special passenger flights over the hump from Chabua. Actually, five C-46s were assigned to the Assam Wing Headquarters at the 1333rd AAFBU to support the special flights. Four planes made the trip, with one aircraft held in reserve.

A Priority & Traffic meeting was led by Major MacGregor, Division Headquarters, with representatives Lt. Wyatt and Capt. Phillips from Washington, D.C., to discuss a variety of topics, including the use of Black troops for the loading and unloading of aircraft. It was decided, instead that Indian Pioneers troops were to be used immediately. Representatives from all bases concurred.

As 1945 continued, more changes were made to ensure the safety of the crews and to increase the tonnage of goods to China. In April the positions of Air Traffic Control Officer and Wing Weather Officer were added to Wing Ops at Chabua, with a resulting decrease in accidents. Teletype transmission lines and land lines were added, however, some stations, especially Misamari and Sookerating experienced problems.

Once the Weather officers attended classes for pilots, they then became the qualified instructors for the pilots, thus freeing up the forecasters. In April there were 33 accidents:
- Eight in flight
- Four on takeoff
- Three missing
- One on landing

Mohanbari had the best accident record in March and the worst record in April, when five C-46s were destroyed and one other suffered major damage. Chabua had three C-46s

washed out, and two others had major damage. Misamari reported two C-46s destroyed, one C-46 with major damage, and one C-47 had minor damage. Tezpur and Jorhat each lost just one plane each—a C-109 and a C-87. There were three fatalities and ten men missing as a result of these accidents. Causes reported as:
- Pilot error—11 (Considered preventable)
- Structural failure—8
- Power plant—7
- Weather—1
- Unknown—6

9.3 PRIORITIES AND TRAFFIC

Along with shortages of parts and supplies, there was a shortage of processes and procedures. No Standard of Operations (SOPs) existed. What was done in Europe did not fit CBI.

The 1943 tonnage went from 987.884 tons in October to 2,912.080 tons by December, and by July 1945 the monthly total was an amazing 71,043 tons. The Priorities and Traffic section (P&T) had developed turn-around procedures, which were responsible for significant tonnage increases in three short months. [1]

On 13 February 1945, Col. George D. Campbell, Jr, West Point 1932, assumed command of HQS Assam Wing Command, replacing Col. Baker, who was rotating stateside. A senior pilot, Col. Campbell had served as Assistant Chief of Staff –Operations, Pacific Division, ATC. The weather during the first part of February was sporadic. After the 15th, good weather prevailed. As a result of the improved weather there was a substantial increase in flight activity over January. In January 1,200 flight plans were handled, with 600 inflight clearances issued. In February that increased to over 28,573 flight plans, with 16,068 inflight clearances handled by Chabua.

Icing created the worst situations, making standard instrument clearances impractical, and necessitating new clearances thought up on the spur-of-the-moment to expedite ship movement through icing conditions.

In May weather conditions at Ledo were closed in. The requested IFR let-down and take-off was disapproved because of conflicting radio band probabilities. A minimum ceiling was prescribed. C-46 operations into Ledo were condemned by the 10th AF, and ATC made the same ruling.

9.4 AIR CORPS SUPPLY

Requisitions for supplies for China bases were submitted to Wing Headquarters Operations, Air Corps Supply, at Chabua, then were sent on to Calcutta, the supply source. With 16 officers and 20 enlisted assigned, the understaffed Air Supply Corps unit attempted to trace lost supply requisitions from China. Many of the requisitions submitted were never received by at Chabua. The trace never revealed what happened to the missing requisitions or the supplies ordered. A new method suggested by China eliminated the middle man at Chabua and had the requisitions sent directly to the source of supply at Calcutta. The new procedure was implemented over a 60-day period in March-April 1945. [7]

9.5 OPERATIONS

9.5.1   YOKE

In support of Stilwell's offensive in late April 1944, YOKE forces initiated a two-pronged attack south from Ledo,

Assam, to Myitkynia, Burma, and west from Yunnan province in China. ATC moved 8,000 Chinese troops from Yunnan to Sookerating. Hump tonnage was reduced during the operation because of flight delays due to the flight diversion to Yunnan on the homebound leg to pick up troops. [9]

9.5.2   OPERATION ICHIGO

Operation Ichigo was the main thrust of the Imperial Japanese Army in China. It was meant to cut off the coast completely and capture American bases

The Chinese did little in response to address the invasion, expecting the U.S. to deliver them. Mountbatten had his hand full in Burma and wanted help as well. The Chinese saw no need to support the fight in Burma and resented the fact that Stilwell had two divisions of Chinese soldiers postured on the China-Burma border at Yunnan—(Y) YOKE forces.

The Japanese were allowed to march through China with little resistance, attaining their goal of occupying French Indochina, while the Generalissimo waited for the Americans to take care of his problems. The Allied airbase assets were moved in knee-jerk reaction movements.

In late 1944, the largest Japanese force pressed south from the Yellow River, connecting Japanese bases in Manchuria with forces in French Indo-China. The Japanese began to overrun major bases of the 14th Air Force in China resulting in:

- Crumbling Chinese resistance
- Setting in of winter and regrouping of Chinese troops halted Japanese advance before Chungking and Kunming attacked
- Threatening of ATC mission of "Aid to China"

- Delivering more supplies to Chennault and the 14th Air Force
- Moving Chinese troops from the Yellow River.

The troops were moved south to defend the southern bases necessary for attack against the Japanese homeland and bases in Indo-China.

9.5.3 RETREAT

How does an army complete a retreat in front of the enemy? In the CBI Theater, it was achieved by airlift. The Japanese, when taking an Allied airfield, found it empty of equipment and a demolished airstrip. When the evacuation of the airbase at Kweilin was ordered, the ATC was standing by ready to load every ton of bombs, equipment, gas, spare parts, and repair shops flown from India to Kunming and Kweilin and evacuate them.

9.5.4 OPERATION GRUBWORM

Completed one month from its start (Dec. 5, 1944-Jan. 5, 1945), Operation Grubworm, an airlift-within-an-airlift, moved by way of Myitkynia to defend the airfields around Kunming, the
- Chinese 14th and 22nd Divisions
    - 25,000 Chinese soldiers
    - 1,596 animals
- Chinese Sixth Army Headquarters
- Armament
    - 48 howitzers
    - 48 heavy mortars
    - 48 antitank guns
- Signal company
    - 42 jeeps

- Two portable surgical hospitals

The airlift required 1,328 transport sorties, with ATC supplying 597 of those flights. [8]

As the different commanders reviewed situations, troops and matériel were moved as needed. Early in the Hump operations, with four very different viewpoints, personalities and concepts of their roles, difficulties arose. As the operations continued, some spots were smoothed out. Chennault was no longer employed by the Chinese. As commander of the 14th AAF and a USA asset, he still had his differences with General Stilwell.

9.5.5 GALAHAD

GALAHAD, better known as Merrill's Marauders, was one part of Stilwell's campaign to drive the Japanese to the south in Burma. The operation meant that supplies needed by Chennault would be taken away from the 14th AAF. It was a double-edged sword. The campaign would allow more aerodromes to be built in northern Burma, letting the ATC planes fly farther south over the lower spine of the Himalayas. The new aerodromes could be used as refueling stops for the ATC transports. They would have a safer route without the danger of Japanese marauders attacking the slow-moving, unarmed, and unescorted planes. The lower altitude would ensure more planes would get through to the China bases. Theoretically, more supplies could be delivered to the 14th. However, until the operation was completed and the aerodromes were constructed, Chennault and the Chinese would see a decrease in supplies. Chennault criticized the effort, while Generals Earl Hoag and Tom Hardin supported Stilwell.[18] Operation GALAHAD went forward in 1944.

Assam Trucking Company

ATC played a vital role in airlifting thousands of men, mules, jeeps, guns, and equipment of the Fifth Indian Division by C-47 to the front in Burma. What would have taken months was completed in a few hours.[10]

9.5.6 ROOSTER MOVEMENT

In response to Special Orders #110 dated 20 April 1945, Capt. Robert C. Roberts, Jr., was made commanding officer and Operations Officer of a Special Mission Detachment from AAFBU 1328th. At the time the Japanese were moving toward the 14th Air Force base at Chihkiang, China. Designated the ROOSTER Movement, Division Headquarters ordered the transfer of sufficient aircraft and personnel from Misamari to transport the 14th and 22nd Divisions of the Chinese 6th Army from Chanyi to Chihkiang—another airlift within-an-airlift. Despite scheduled 50- and 100-hour maintenance checks, Misamari continued to supply the China Wing with an average of ten C-46s for the mission's duration.

Seventy-seven enlisted men, including radio operators, aircraft mechanics, P&T loading supervisors and Air Corps Supply personnel, were assigned as well as thirty-seven officers, 2nd/Lt Charles Walker was in charge of P&T, plus thirty-five flying officers. Capt. Roberts coordinated all Detachment activities at the directive and in support of the objectives of the Commanding Officer, China Wing and his representatives.

Operations and Engineering functions were located at Kunming. The first three days of functions at Kunming proved difficult until vehicles required for transport of cargo and troops to the planes at Kunming arrived. After a briefing, the aircrews flew empty transports to Chanyi, where they were loaded with forty-four troops, their equipment and personal weapons while the aircrew received a supplemental briefing. The three-hour trip to Chihkiang was usually made on instruments. While the base at Chihkiang had good facilities, the checkpoints at Tushan and Kweiyang did not, resulting in a 327-mile flight without benefit of any signals. They landing strip at Chihkiang, composed of dirt and crushed rock, was about 4,000 feet long.

2nd/Lt. Walker, in charge of the 1328th P&T detachment, working with S/Sgt. Robert B. Collen and Pfc. James F. Rierson, were responsible to ensure that planes were properly configured with stables for the horses and donkeys. Walker then worked with three enlisted from Chanyi P&T directly under Captain Boedecker, Director at Chanyi, a 24-hour exercise with two other officers, one from Chabua and one from Mohanbari, in charge of loading equipment. There were many challenges.

Lt. Walker drew up loading plans on the spot, and with the help of enlisted men supervised the loading of each ship. Three Clark lifts were employed for the loading of large equipment. Unlike some vehicles that could be disassembled, the 37 MM anti-tank guns had to be loaded in one piece. To keep the weapon from toppling over in flight, Chinese soldiers were allowed to ride the weapon on the lift to aid in balancing the gun while it was being moved to its tie-down area. The 75 MM Pack Howitzers were loaded best in one piece as well. Once the procedures were developed on the fly, loading went smoothly, until Chinese soldiers refused to allow Chinese laborers to handle ammunition. The Chinese soldiers, who proved to be well-disciplined, were allowed to load their own ammunition.

P&T personnel were required to annotate important details. A ship was not overloaded and weight and balance forms were submitted. The Commandos carried 46,000

Assam Trucking Company

pounds gross take-off weight, including forty-four infantry troops, their weapons, plus 1,100 gallons of gasoline for fuel. The weight of each troop with equipment was arbitrarily set at 186 pounds.

The U.S. 19th Veterinary Corps were essential and efficient in the handling and loading of horses and donkeys. Special configurations constructed in the C-46 ensured that the animals' movements were limited. Each animal had a Chinese handler assigned to keep the animals calm. The animals usually weighed in at 1,000 pounds each, 754 pounds being allowed for the horse, handler and baggage. The final payload averaged about 4,744 pounds per ship. [11]

The Rooster Movement turned out to be the most efficient and successful large-scale air transport mission ever attempted in the India China Division. A total of 25,136 troops, 2,178 horses and 1,565 tons of cargo were lifted by the Air Transport Command.

Chinese soldiers or refugees were often moved within China as needed to keep them out of the way of advancing Japanese.

Chapter 10.0 — Search & Rescue/Medevac

Unlike the other two theaters in the war, CBI lost planes that were not behind enemy lines or over open ocean, leaving those in the theater open to create new programs of operation. Search and Rescue was one of those services.

10.1 AAFBU 1352nd SEARCH AND RESCUE

Begun as an impromptu group in response to the crash of C-46 Flight 12420 with 18 military and civilian passengers, including well known journalist Eric Sevareid, Search and Rescue, better known as "Blackie's Gang" for Capt. John L. "Blackie" Porter, was supported with those available and who could be spared from other duties. They would do aerial searches for planes and crews reported downed or missing. Officially begun at Chabua in July 1943, Porter was assigned two C-47s that were armed with two .30 caliber guns each for protection. One was fired by the co-pilot, who held the gun in his lap and shot from the pilot compartment window: the

Assam Trucking Company

other was manned by other crewmembers, who fired from the cargo door. Their mission was to locate downed aircraft and report same. If possible, they were to also render aid in the form of air-dropped medical and food supplies, and to guide ground rescue parties to the aircrew.

In October 1943, Porter was named Flying Safety and Rescue Officer at Chabua. Then on December 10, 1943, Blackie's Gang flew their last mission. While preparing to go on a drop-mission, they heard a May-Day call from a C-47 from Chabua which had made an emergency landing at Ft. Hertz. Enemy aircraft activity was reported in the area. When Porter and his aircrew arrived overhead, they saw that the C-47 was on fire, but the crew had safely evacuated the aircraft. In the meantime, Porter's plane was attacked by Japanese fighters, who shot out one of their engines. One man, James Spain, was able to jump out, but Porter stayed with the plane, trying to maintain sufficient altitude to allow the others on the crew to jump clear. Those who lost their lives with Capt. Porter included:

- Walter R. Oswalt, Radio Operator
- Harry D. Tucker, Gunner
- Harold W. Neibler, Ariel Engineer [1]

The 1352nd Search and Rescue AAFBU was organized at Chabua in February 1944. In January 1945, Lt. Col. Gordon Rust, Division Integration and Support, arrived to make the final plans to move Search and Rescue 1352nd AAFBU to Mohanbari AAFBU. Their mission was expanded to provide

- Air search and rescue for the entire Hump route network flown by all theater forces.
- Ground net of contact and escape route network, encompassing all friendly forces, allies, and natives of the area in CBI

- Flying safety information to pilots and aircrew members necessary to prepare them for survival and ground rescue in the jungles, mountains, and widely varying climatic conditions encountered on theater air routes
- Return the bodies and personal effects of those who died in crashes on the Hump

Major Donald C. Pricer, commander of the 1352nd, had hand-picked the personnel for reassignment from other ATC units, where they were qualified in the unit assigned aircraft and the ATC transport mission and were familiar with the CBI area topography and weather. Flight Operations were meticulous, and required precision and knowledge of weather conditions as well as the capabilities of the aircraft and aircrews.

Like the aircrews, the maintenance personnel were hand-picked to maintain "in-commission" status on nine different aircraft types and models [C-47, B-25s, (Models H, G, J), U-64 Nordsman, L1, Piper L-4s, Stinson L-5s, and Sikorsky R-4s]—single and twin engine, frame and canvas to sheet metal, and precision instrument repair. The Willis jeep, communications units and GMC amphibious Duck were used as ground equipment. The parachute section was capable of repacking of aircrew parachutes, jump chutes and airdrop chutes of all sizes. The Unit Administrative Staff, Headquarters Staff and Supply personnel, like the others, were hand-picked based on the qualifications and ability to provide the necessary services to the unit. The Intelligence and Rescue Coordination and Control Staff took charge of all activities for all flight crews and ground search personnel. The most unique were the Flight Surgeon (Jump Qualified) and Para-Medic Team, and the "Jungle Wallas," a select group of outdoorsmen, loggers, and roughnecks who were Jump-

Assam Trucking Company

Qualified at the British Para-Training School at Rawalpindi, India.

Maj. Pricer had a large map in his operations office. As the planes flew the Hump in either direction, they reported their positions at stated intervals. When a plane didn't come in, a colored pin was inserted in the map where it was last heard from, providing a triangulation point. The Search and Rescue team would go out looking for a spiral of smoke, a slash of timber, the blink of a mirror—anything to spot the scene of the crash.

Their operating procedures were to
1. Find the crash site
2. Establish a set of working signals with the crew to find out how many were dead and hurt, what supplies were needed, and to assure them that help was coming
3. Carry out rescue

If a flat patch was located close to the site, a Search and Rescue Unit (SRU) might land a small plane and fly the aircrew members out. If no flat patch existed, the SRU would drop a jungle-wise team with medicines, supplies and litters, and walk them back to civilization through the jungle. Some crash crews made it back on their own, dodging dangerous snakes, tribes identified as headhunters, living on berries and tree bark and mapping the area as they made their way back through the maze of mountain ranges.

Pilot Charles G. Allison and his crew of three held the long-term record for coming out alone. They emerged after 93 days, barefoot and beaten. The Army doctor pronounced them in fit condition. One of the crew members noted it was "a hell of a way to get in shape."

In February 1945, an S&R mission was sent to the aid of a Chabua-based C-47 that had crash-landed near Santahar in

Northern Bengal. The C-47 was transporting 19 Chinese mental patients and a few Americans back to India from a combat area in Burma. The crash was caused when one of the Chinese dropped a phosphorus grenade he had smuggled on board the aircraft. The grenade exploded, causing the plane to be filled with flames and fumes, temporarily blinding the pilot. Maj. Robert J. Seabolt managed to land the aircraft with no injuries to passengers or crew. The only casualties were those on the ground—five head of cattle. Chowkidars (local police) investigated the crash. [2]

No matter how long it took for the downed airmen to walk out or be rescued, no lost plane was given up or struck off the list until every man who was aboard the ship was fully accounted for. In 1943, 2% of personnel missing were rescued, and during the first 6 months of 1944 77% were rescued. When the unit was shut down in December 1945, only 33 unknowns remained of the original 893 unknowns (when the unit was first organized in 1943) and 15 of those still on the books were located at altitudes and isolated to risk the loss of other lives to recover the bodies and personal effects. [3]

10.2 AIRCRAFT

10.2.1 L-1 Stinson

A single-engine monoplane with limited range. Only two were assigned to S & R. Modifications included the installation of a 30-gallon fuel cabin tank in place of the back seat and a deck over the tank, allowing ambulance capability for the first time. The L-1 was unique in that level flight could be maintained at 25 mph.

Assam Trucking Company

10.2.2 L-4 Piper

The unit possessed two L-4s, which spent most of their time on the ground. They were grossly under-powered, with takeoff distance greater than acceptable. The L-5 Stinson was more efficient.

10.2.3 L-5 Stinson

The unit include five of these fine little planes, considered the workhorse of the small aircraft. Their speed, maneuverability, a reasonable service ceiling and limited instruments for instrument flying-range, plus a rear seat conversion to an ambulance aircraft, permitted all the versatility lacking in the other small aircraft available. Short-field takeoffs and landings with brakes set were often required when maneuvering in mountain and jungle "do-it-yourself" landing strips, or on river bars, and using the Ledo Road as pickup points.

10.2.4 UC-64 Canadian Norseman

Built to U. S. Army Air Corps specification, two of these aircraft were received in 1945. The requirements included increased fuel capacity, radio equipment, etc., and reduced payload when fully fueled. The major weakness for short field and road operation of
this aircraft was its minimal wheel braking that resulted in ground looping on takeoff. This one limitation severely affected operational use for rescue.

10.2.5 R-4 Sikorsky Helicopter

Two helicopters were received by the unit in 1945, shipped from the U. S. by C-54 partially disassembled.

Assembly, testing, and obtaining a TDY pilot consumed considerable time. These aircraft were not used on rescue missions, as the bulk of the missions following their operation readiness did not require the helicopter's talent.

As with most things on the Hump, the aircraft used by Search and Rescue were late in arriving and not fully tested for their operational requirements. Time, lack of trained personnel, testing for operational requirements, modifications required and limitations of the aircraft for mountain flying resulted in most of the aircraft being used on a limited basis, or not at all.

10.3 CRASHES & BAILOUTS

The last commander of the ATC in CBI, Gen. William Tunner, said the ideal pilot to fly the Hump, the world's most dangerous flight, should have the following qualifications:
- Age: 30s
- Pilot time: 2,000 plus hours
- Pilot time over the Hump and/or India: 200 hours
- Monthly Pilot time: 30 plus hours
- Being crazy helped too

Most of the men flying the Hump did not fit these qualifications, especially toward the last year of hostilities. They were young and, for the most part, just out of Aviation Cadets, and looking for adventure.

No matter how mature or experienced the aircrews were, accidents would and did happen. With all the problems they faced on a daily basis, it was a wonder how many finished their missions with few or no negative events.

10.3.1 December 1943

As an inexperienced Hump pilot, Kenneth G. Framsted, had a memorable December. He made his first Hump run on the 17th, to Yunnanyi Air Field. Upon landing, he headed for the snack bar and ordered the fresh eggs he hadn't had since leaving the States three months before. The food rations at Mohanbari consisted of powdered eggs, powdered milk, Spam and marmalade, which grew old very fast. Inside he saddled up on a high stool and placed his order with a slight Chinese fry cook who asked, "How you like your eggises—hard fly, soft fly or sclamble?" He ordered six eggs "soft fly. He was pleasantly surprised to see the eggs came with home-fried potatoes, toast and a mug of hot tea.

The second memorable day began sometime after midnight on Christmas Eve. The mission was again to Yunnanyi. The takeoff was made in total darkness, and the landing just after sunup. Landing to the south from a left-hand pattern, the sun poured into the cockpit from the copilot side of the aircraft. At the appropriate time for lowering the landing gear, upon command of the pilot, he did so. They checked the gear-locked position lights on the instrument panel and making a visual check as well. All seemed in order. Flaps were lowered, the approach continued, and the runway was coming up in fine shape. As the power was reduced just prior to flare-out, the one sound a pilot does not want to hear went off the—"Gear Unsafe." horn blasted in the cockpit. The pilot reached up and flicked a switch deactivating the horn and continued a normal touchdown. Normal, except that power was not added to affect a go-around, which was the accepted SOP because a touch-down should lock the gear in the down position. The procedure was to become airborne, make a go-around (known as a touch-and-go), and double check for a positive gear down and locked condition. The next thing they knew they had a "sinking feeling" as the plane began settling onto the runway. The props hit first, digging into the red clay. There was nothing left to do but cut the ignition switches to prevent a possible fire and wait for the long slide of metal against earth and rock to end. When it did, the crew evacuated by way of the cockpit emergency door. No one was hurt, the runway was blocked for most of the day as the plane was unloaded of salvageable items, and the aircraft was positioned along the flight line, becoming the famous (or infamous) snack-bar of Yunnanyi.

With limited communications capability at the time, no word was passed back to Mohanbari of the problem encountered. The crew caught the first available flight back.

When they walked into their Base Operations that night with other crew members for debriefing, the base administration officer was there, surprised and glad to see them. He informed the crew that he had their MIA report made out, but had not as yet transmitted it to the War Department. [4]

10.3.2   March 1944

As with most things, Murphy's Law prevailed, and it was no different for the 1352nd AAFBU Search and Rescue. During the month, at the request of the 1352nd AAFBU Rescue Unit, three separate attempts were made to reach the wreck of an airplane which crashed north of Misamari. Success favored the third attempt and the ship was identified as C-47 #42-100614, reported missing by the Rescue Unit on May 25, 1944, on a flight from Sylhet to Dinjan.

The first party, consisting of four officers, one enlisted man, seven Gurkhas, and fifteen Duffla porters, left the base

Assam Trucking Company

on March 10th, and after four days was forced to return to base. A Handy-talkie, which would have been of invaluable use in maintaining contact with an aerial observer, was dropped to the first team, but was smashed when its parachute broke away from the package in mid-air.

The second party left on March 18th traveling much more lightly. This party consisted of four officers and two Duffla porters. Although successful in reaching the close vicinity of the wreck, (75 yards from it as advised by aerial observers), this party failed to find the wreck in the dense underbrush.

The Base Commander, Lt. Colonel Pratt, had been convinced from the start that, judging from landmarks discovered from the air, he could go directly to the wreck. At 0300 on March 24th, Col. Pratt and Maj. W. H. Volz, Director of Supply and Service, left the base on a third attempt. They were successful in reaching the wreck and brought back information establishing the identity of the aircraft. [5]

Many planes were lost, some with loss of lives and some with all surviving, by other air units in CBI. The following events were recorded, just during 1945, and only for the ATC units at the bases mentioned.

10.3.3   1945

10.3.3.1 October 1944-February 1945

Robert M (Pete) Loving, Jr., was one of those young men, just 21 when he graduated from Av Cadets and 23 when he reached to India. He was based at Chabua. On one of his flights as a copilot with pilot Capt. Whitehurst on a night trip to Kunming, all went well until an hour out, when the oil quantity gauge for the right engine dropped to zero and the oil pressure dropped immediately. The engine feathered, they made a 180° turn for home, but couldn't maintain their altitude. The pilot decided it was time to jettison the cargo in order to clear the first ridge of the Himalayas on their back to base. Pete and the R/O strapped on their chutes and grabbed walk-around bottles (oxygen) then started untying and pushing .50 caliber ammo out the back door. After recharging the bottles many times, Capt. Whitehurst indicated he was able to hold altitude. Exhausted, he climbed back into the copilot's seat just in time to see by moonlight that they were going thru a low place in the first ridge. Before let-down at Chabua, Capt. Whitehurst decided to restart the engine rather than make a single-engine landing. The landing was good and engine shut down as soon as they cleared the runway. Later inspection revealed that the oil sump plug had not been safetied in place after the oil change prior to the flight.

Pete Loving's experiences ended with a landing the crew could walk away from.

Chabua had 34 accidents in early 1945. Nineteen planes were washed-out, nine had major damage and another six sustained minor damage. [7]

10.3.4 JANUARY

Balipara Frontier Tract in the Duffla Hills, part of the Himalaya foothills, was a political "No-Man's" land. No one was permitted to enter the area without permission of the Political Officer of the District tract and Duffla Hills, lying only a few miles north of the base. Bn. Commander Lt. Col. Richard Booth, provided outposts as the Political Officer deemed necessary. Search and rescue parties with patrols were allowed to penetrate the Duffla Hills to an area about eight days march from the edge of the valley, crossing the Kukrees, symbolic of the Gurkha military units who aided the

Misamari S&R parties.[8] The Gurkha were an indigenous tribe of Nepal known for their ferocity in battle and the slightly curved and very sharp knives they carried.

The Maharajah of Nepal had prohibited flights over his country. The people believed it angered the "gods" if mortals violated their domain. The same prohibition of the airspace over the tiny country of Sikkim, sandwiched between Bhutan to the east and Nepal to the west, existed as well. [9]

10.3.5　February 1945

C-46, #6637, piloted by Capt. Frederick J. Telecky with copilot 1/Lt. Gordon Malone and R/O Sgt George B. McBride, left Yunnanyi on 2/12/45. At 17,500 they hit turbulence over Yunlung. Making a 180° turn to escape the turbulence they hit a downdraft and descended to 14,000 ft. The copilot hit the blowers, only to have both engines cut out. The plane was still descending when the R/O and the copilot bailed out at 13,000 ft. followed by the pilot at 12,000 ft. Capt. Telecky watched as the plane crashed and burned below him.

Once on the ground, the pilot and copilot joined up, but couldn't find the R/O. They walked to a village they had noticed in the valley below. At first the villagers were afraid, but then invited the crew in and fed them. The villagers organized a search party to help find Sgt. McBride. He was found unharmed in a nearby village. They waited two days for horses that had been requested until help and the horses arrived. Unable to ride the horses, they walked back to Yunlung, a distance of 30 miles. It took them two days to make the journey on foot. One day later a rescue mission arrived. The Chinese were pleased when they were given rupees and gifts for helping. The next day, with the rescue group, they walked forty miles to the Burma Road, then stayed overnight with an Ack Ack group. On the sixth day they were driven by truck to Yunnan and spent the seventh and eighth days in a hospital there. [10]

There were eight crashes noted of aircraft from Sookerating. Six C-46s washed-out, one sustained major damage and one had minor loss.

10.3.6   MARCH-JUNE

Not everyone was so lucky. C-46, tail number 1045, crashed near Kunming on March 25, 1945. The crew included F/O Donald W. Chisholm, Pilot; 2nd Lt. Kenneth E. Thompson, Copilot; and Pvt. Clifford W. Moore, Radio Operator. They departed Misamari on March 25th at about 2100 IST for Kunming. The weather was poor from the outset making it necessary for them to climb to 19,000 feet to get over the first ridge. As a consequence, they burned a great amount of fuel. Their radio-compass indicator was spinning, and they were given a bad bearing from "Droopy Homer," which caused a loss of half an hour. A genuine bearing later from Kunming found #1045 about 160 miles north of Kunming. The pilot turned on a bearing of 225° and flew for an hour. At the time, when they were supposedly over Kunming, no contact could be made. It was determined that they were lost.

Short of fuel and unable to obtain bearings because of several other Maydays already on the air, they advised Kunming that they were going to jump, then they flew west for half an hour. Fuel supplies almost exhausted, they jumped, with the copilot first, R/O, second and pilot, third. Despite the fact that the ship was heading west and the copilot jumped first, both the pilot and R/O landed east of the copilot. The R/O was the first to reach a Chinese village. The crew had not been able to immediately join-up after bailing out. With the help of a Chinese search party the R/O was able to find the copilot. The pilot could not be found, but later they heard he was in a neighboring village and that his ribs were fractured.

Remaining in the village that night, they were fed and treated well. They were offered ten Chinese women for the evening's pleasure. Both men refused as politely as possible.

The next morning, they mounted horses for the journey to the village where the pilot was located. A man who was acting as boss in the village turned up in the uniform of a Chinese major, leading about twenty guerilla soldiers. As the party proceeded down the village street, school children of the village cheered, and one blew a bugle.

Escorted by the soldiers, the men rode to the other village and joined the pilot. They taped his cracked ribs and made plans for the journey out. The Chinese again led the party and took them to the base at Luliang. At the beginning of this trip, they did not know where they were being taken. They reached Luliang at about 2200 hours on March 27th, when they contacted Operations to report their situation.

The pilot suggested that a decent English-Chinese "Pointee-Talkie" be included in the jungle kit rather than the phonetic sentences, which the Chinese only laughed at.

A puzzling problem in Hump flying was centered around the crew of Misamari C-47, tail number 738. The first information of their distress was contained in a wire from Chanyi on March 24th. The flight left Misamari at 1420 IST on March 23, 1945, en route to Chengkung.

The crew of #738 was believed to have bailed out at about 1600 GMT, last contact being Kunming at 1552 GMT. Bearings taken were:
- 206° by Kunming at 1505 GMT

- 265° by Yunnanyi at 1510 GMT
- Reciprocal bearing of 267° by Chanyi at 1514 GMT.

A triangulation of bearings indicated the location was just northeast of Yangki. One wire was received on #738, which was later cancelled and filed.

The next information submitted to Misamari was four days later, on March 28th, which gave the position and condition of the crew in a wire from Kunming.

Crew of #738 were safe about 94 miles southeast of Kweiyang, and were heading for the nearest American base. (This position placed the crew near Japanese lines.)

The puzzle was in the fact that, despite the encouraging news, no further information was received by the end of March. Word of the crew was received in April when the pilot, 2/Lt. Charles O. Foster; copilot 2/Lt. James H. Vogel; and, R/O S/Sgt James H. Blue, walked into Kunming. The cause of the bailout was determined to be that they were lost and ran out of fuel.

On June 27, 1945, Misamari suffered its first fatal accident since March 9th, when a ship became missing with four men onboard. The unit had flown 45,000 hours, some 7,200,000 miles without a fatality. The crewmembers of this ship, C-46 #3615, were:

- Pilot: F/O Harry S. Roberts
- Copilot: F/O Maynard B. Long
- R/O: Pfc Arville M. Mooney

Word was received about the location of a wrecked aircraft from tea planters who lived near Bhutan. Two wires indicated that the crash occurred on June 5, 1945, at 2145 hours. The plane had actually been seen as it burned on a incline. The wreck was located at 26° 57' North, 91° 54' East at approximately 5,000 feet up on the side of a steep hillside.

Assam Trucking Company

An aerial sighting was made on June 8th from a PT-17, and later more sightings and aerial photographs were made. Through the trips and photographs, a broad field in front of a Planter's Club, about four miles north of the village of Paneri was identified as a suitable landing field for the small plane. On the afternoon of the 8th, two planes landed on the field and gained permission to use the location for a search party headquarters, and to visit the wreck. This was necessary, as the crash occurred in Bhutan, an independent state. Permission was granted by Mr. Proqhan, Bhutan Officer and representative of the British government, who volunteered the service of porters and guides from his own village. The 1328th scarch party met the porters and guides at the Corramore Tea Estate on June 17th, and reached the crash site on June 18th after a difficult climb. The plane had obviously burned intensely for quite some time. No bodies were completely whole, making it difficult to determine how many people were on board. One estimate was that nine bodies were found. After the remains were buried, the graves marked, and the I.F.F. destroyed, the wreck was marked with a large yellow panel to prevent any later unnecessary visits to the same wreck. [13]

10.3.7 APRIL

During the month of April, the base Intelligence and Security (I&S) officer, Lt. Clovis C. V. Cailliez, and Capt. Keith D. Swisher, Base Search and Rescue Officer, maintained close liaison with the Search and Rescue Officer at Mohanbari. A ground party was organized and initiated a request at Jorhat from the Mohanbari Search and Rescue unit to investigate a wrecked plane reported near the village of Shakchi, Burma. An IFF plate brought out by the natives and taken to

Mohanbari was checked against the records of the communications department. All indications were that its identity was aircraft #253, missing from Jorhat. Later during the month, at the request of Mohanbari AAF Search and Rescue Squadron, Lt. Cailliez arranged with the Quartermaster Graves Registration Unit to bring back the bodies of Lalmanir Hat aircraft #747 from Monchen to Jorhat for burial in the military cemetery at Jorhat. Sgt Adams from Graves Registration returned to Jorhat with a 17-year old Naga boy, the son of the village chief, Umpi. Since he had proven English skills, he served as an interpreter, on all Search and Rescue missions from Jorhat. [15]

In April five C-46s crashes occurred, with no crew losses.

#7165 crashed 20 miles SE of Mohanbari after experiencing mechanical failure of the blower section in the left engine. Crew on board
- 1/lt. Joe Walters, pilot
- F/O Howard Turpin, copilot
- Sgt Tony Loza, R/O
- Sgt John W. Hancock, Engineer

#4750 crashed 22 miles west of Mohanbari when the left engine failed. Crew on board
- 1/Lt William F. Vickers, pilot
- F/O Emerald C. Case, copilot
- T/Sgt Richard M. Harding, R/O.

#1153 crashed two miles from base three minutes after takeoff, with the loss of the left engine. Crew on board
- F/O Joseph M. Lambert, pilot
- F/O Warren W. Clements
- T/Sgt M. L. Shapiro

#1053 crashed 70 miles SE of the base when they lost both engines. Crew on board
- 2/Lt Peter R. DeLonza, pilot

- 2/Lt Donald W. Prindle, copilot
- Pfc. Robert H. Duke, R/O

#6984 crashed on takeoff from Chanyi due to pilot error. They had not removed the elevator locks prior to takeoff. Crew on board
- 2/Lt Andrew J. O'Neal, pilot
- F/O Jason O. Hine, copilot
- Sgt George J. Munas, R/O

In May 1945 there were two C-46 and crew losses.

#7161A crashed 50 miles from Shingbwjang due to loss of both engines—broken fuel lines. Crew on board
- 2/Lt L. E. Vest, pilot
- 1/Lt Harley L. Tracy, copilot
- T/Sgt Wilfred S. Almond, R/O
- T/Sgt Roy T. Brown, Crew Chief (C/C)

#4749 with loss of crew and aircraft in flight, cause unknown. The plane and crew were never found. Crew on board
- F/O Max Gillespie, pilot
- F/O Stephen B. Pond, Jr., copilot
- T/Sgt. F. A. Radics, R/O

10.3.8 JUNE

In June 1945, twenty-four accidents were reported by Chabua, Mohanbari, Jorhat, Misamari and Sookerating. Of those recorded,
- Fifteen were destroyed,
- Six sustained major damage and
- Three had minor losses.

Causes for the crashes included mechanical failure, pilot error, and three ran out of fuel.

10.3.9 AUGUST

Flight 23 from Calcutta to Chabua crashed at Lalmanir Hat on August 5th after overshooting the runway on an attempted landing. All on board, 4 crew and 16 passengers, perished. The crew was from the ATCH base at Dum Dum. Passengers assigned to Lalmanir Hat returning from TDY in Calcutta included Capt. John T. Billingsley, 903rd Veterinarian Food Inspection Detachment, Calcutta and Pvt. Johnnie L. Woodard, MP.

On the 7th of August a C-47 en route from Ondal to Tezpur crashed 20 miles from Lalmanir Hat near Rangpur. A 12-member rescue team was sent. The team traveled by rail, truck, jeep, and finally "weasel" supplied by the Signal Corp unit. The lesson learned from this experience was that there was no adequate vehicle assigned to the base for rescue operations in tropical terrain. It was recommended that a weasel be assigned to every base and properly equipped with emergency medical and rescue gear. Ambulances and/or jeeps could not be taken off base, especially in monsoon season, because they would bog down in the mud and water. [16]

10.4 ACCIDENTS—January 1945

January 6-7, 1945, there were thirteen fatalities, nineteen missing including nine passengers on a C-46 from Chabua in January. The worst ever weather was experienced during these dates. [6]

Assam Trucking Company

	C-46	C-87	C47	B-24D	C-109	TOTAL
TOTAL WRECKS	5	1	1	0	1	8
MAJOR DAMAGE	3	2	1	1	0	7
MINOR DAMAGE	0	3	0	0	0	3
MISSING	3	0	0	0	0	3
TOTAL	11	6	2	1	1	21

TABLE 10-1, January 1945

	WASHOUT	MAJOR DAMAGE	MINOR DAMAGE	TOTAL
CHABUA	3	2	0	5
SOOKERATING	4	0	0	4
JORHAT	0	2	2	4
TEZPUR	2	1	1	4
MOHANBARI	1	2	0	3
MISAMARI	1	0	0	1
	11	7	3	21

TABLE 10-2, Breakdown of Accidents by Base unit

CAUSE	SOOK	JORHAT	CHABUA	TEZ	MOHAN	MIS	TOTAL
Pilot Error	0	2	1	1	2	1	7
Mech Fail	1	1	2	1	1	0	6
WX	2	0	1	1	0	0	4
Unknown	0	0	1	1	0	0	2
Fail-Rnwy lights	1	0	0	0	0	0	1
Bird strike	0	1	0	0	0	0	1
	4	4	5	4	3	1	21

TABLE 10-3, Breakdown by Cause

10.5 MEDEVAC

While there had been very basic attempts made at the end of WWI to transfer an injured pilot to medical facilities, the concept of full-blown medical evacuation by air was developed and implemented at Bowman Field, KY, as the 621st Air Evacuation Squadron, and activated on 11 November 1942. The 803rd Squadron was created 16 April 1943 under the command of Maj. Morris Kaplan, and vigorous training was begun. By late August 1943 the squadron was moved to Camp Anza, CA. On 7 September 1943 they boarded troop ship George Washington, destination CBI. Two months later they arrived at Calcutta and were flown on to Chabua, which would serve as their base of operations.

The squadron was made of 94 personnel
- 5 flight surgeons

Assam Trucking Company

- 25 flight nurses
- 1 supply officer
- 4 supporting sergeants
- 24 trained medical technicians on flying status as tech. sergeants
- 35 general support troops

The squadron was attached to Headquarters ATC, while the commanding officer was attached to Gen. Stilwell's staff as Theater Air Evacuation Officer. Other Medical Air Evacuation (MAE) Squadrons supported the other AAF units

Most of the nurses had served as stewardesses with American Airlines. In the early days of passenger flight in the United States, women hired as stewardesses were required to be registered nurses. During WWII the requirement was relaxed, as most women with RN training were joining the military. The group was quartered in a tent camp near Chabua, on the Hattialli tea plantation. It was nicknamed "Pantyalley."

Duties began immediately, with flights of patients from the 20th General Hospital at Ledo to Calcutta and Karachi, with some being sent on the Zone of the Interior (ZI—United States).

With stepped-up military action, a flight of one flight surgeon, six flight nurses, and six technicians, was attached to the 14th AAF and scattered over the area to coordinate and carry out evacuations of American and Chinese casualties. They became extremely busy, making two to three flights per week over the Hump to Kunming and 2 to 3 flights per day on the Burma circuit, with 30-40 casualties per load. Several flights were made weekly among the bases in India, gathering American and Indian casualties for transport to various American hospitals and transfer to the United States.

The ATC, 14th AAF and various other combat organizations pressed the Medevac organization into service for numerous emergency rescue missions in China and Burma. Some of the missions took place under very dangerous us conditions. The day after the taking of the air strip at Myitkynia, three of the 803rd personnel, Capt. Louis Collins, Flt. Nurse Audrey Rogers and M/Sgt. Lee Miller, were wounded by a Japanese fighter strafing the field. The Chinese litter patient being loaded was killed instantly by shrapnel. Capt. Collins received shrapnel wounds to his upper arm and hand; Sgt. Miller had shrapnel wounds to his inner right arm and the base of his skull; and, Lt. Roger suffered shrapnel wounds to her right knee and thigh. The patient, who died, was the only death in over 35,000 patients carried the first year.

In the first year there was almost a regular schedule of flights for evacuation within India, to many stops in Burma, and over the Hump to China. Flights became routine, in that the loads of patients became increasingly heavy, with increasingly severely wounded. After the first year, about half the nurses (those who wished to) returned to the States. In December 1944 the rest of the nurses and their commanding officer were returned to the ZI. Dr. Duncan was promoted to Major and became the new CO. New nurses were arriving. Suddenly the patient load decreased, and remained down until the end of hostilities. [13]

"The record of medical service in the early period was one of expediency and frustration in the face of incredible filth, the unconquered specter of malaria, the dependence on British colonial medical facilities, the lack of organization and housekeeping equipment and supplies, plus a feeling of remoteness from the rest of the fighting world." [14]

Chapter 11.0 — Closing Shop— 1945

11.1 AAFBU 1329ated DERAGON

With the end of the war in Europe, focus turned to the Pacific. In accordance with Operations Order #4, dated 28 January 1945, each base started closure procedures when their allocation of aviation gas/fuel had been airlifted to China. Misamari and Sookerating started closure procedures toward the end of September, 1945.

On the 14th and 15th of January 1945, all equipment and personnel were moved in preparation for closing from Deragon to the 1327th at Tezpur, 48 hours ahead of schedule. General Tunner congratulated them for their effort and requested full details of the move to be documented as an SOP. On the 27th of January, the 3rd Air Transport Squadron (ATS) was attached to the 1337 AAFBU, Sookerating and began passenger shuttle line operations to Assam Wing bases and Lalmanir Hat as of 1 February 1945. Deragon closed as an ICD base on 10 June. Misamari, and Sookerating started closure procedures toward the end of September. [1]

11.2   AAFBU 1328th, MISAMARI

Misamari's allocation of 4210 gallons of gas was loaded for delivery to China. When the allocation airlift was completed, deactivation closure procedures began. Misamari delivered their last fuel allocation to China in September 1945. Misamari closed on October 20, 1945.

11.3   AAFBU 1330th JORHAT

At that time, CBI saw an influx of supplies they had ordered months or as much as a year earlier. At Jorhat, a new ice plant was received in June 1945. Quartermaster Corps reported there were no shortages and food was in good supply. While the electrical power was still inadequate, there had been an improvement in the water system.

The last C-87, #43-30589, with a load to China flew out of Jorhat on 20 September 1945. The four-man crew included Captain Robert J. McClurg, pilot; F/O A. E. Saunders, Copilot; Pvt. John J. Goldsmith, R/O; and Pfc. Harvey C. Mason, A. E.

The gathering of war material and machines started. Forty flight crews were selected to ferry consolidated B-24s from Karachi to Shanghai for the Chinese Air Force.

With the surrender of the Japanese, orders to deactivate the Jorhat were received on 19 September, with deactivation scheduled for 11 October. On 26 September 1945, the last operational C-87, #44-39280, departed Jorhat at 1330 for Bangalore, leaving 1 B-25, 1 C-47 and 1 PT-19 still attached to the base. Jorhat began operations on 15 April 1943. During the two-and-one half years of operation, crews from Jorhat made 16,610 trips, carrying 72,226 tons of supplies. The base suffered 149 fatalities—one for every 111 ½ trips to China. [2]

11.4 AAFBU 1327th, TEZPUR

Even with the war taking a turn to the east, life in India went on as before. A smallpox epidemic broke out among the townspeople of Tezpur and Ranjipara North. Medals were given, pictures taken of the awards, new ways of completing maintenance tasks were developed. For instance, Pfc. Roosevelt Moore developed an easier way to remove a tire from a rim. 1/Lt. James L Mayles, Jr., made a diagram of the Hump Radio Chart covering all the Hump Routes, mileage, bearings, radio contacts and central boundaries. The chart was especially useful to the newly-arrived, who were issued a folder for "home" study.

Base security began posting guards on all gas dumps, ammunition storage areas, radio installations and drinking water tanks. The step-up in surveillance was to curb the threat of sabotage.

On 5 March 1945, ICD letter 353/230, Headquarters, ICD-ATC directed the use of more civilian employees where possible. Training programs were established for local men as apprentice airplane mechanics. Major Robert C. Thomas, Base Engineering Officer, supervised the training of forty-five locals who were cleared for classes. Each class consisted of five Indians, with one speaking English. A similar program was begun for African-American personnel, to train them for upper or more complicated maintenance, 1st echelon. The emphasis was to free the Blacks from loading/unloading, etc., and utilize them to the greatest extent possible in more responsible positions according to their qualifications. They were to be assigned as motor pool mechanics, administration clerks, supply and utilities clerks, electricians, carpenters, cooks and airplane frame mechanics, aircraft sheet metal workers, welders, and refueling unit operators. These were skills they could use Stateside.

The last scheduled flight to China was made on 29 May. The base was deactivated on the 31st of May, with all assigned aircraft transferred to other base units in ICD. (Tezpur). With the deactivation of Tezpur came the reduction in the number of aircraft assigned to the Assam Wing. Most of the crew and maintenance personnel were reassigned to Jorhat, Kurmitola and Shamshernager. Administrative personnel were moved to the lower valley of the China Wing.

On the 1st of June 1945 other groups officially came under ATC operational control. On the 8th of June the 443rd Troop Carrier Group at Dinjan, the 7th Bomb Group at Tezgaon and the 308th Bomb Group at Rupsi were assigned to help take up the slack. Due to plane and equipment modification requirements, the ATC mission could not be supported 100%.[3]

11.5 AAFBU 1326th LALMANIR HAT

In May 1945, improvements to the facilities were made to handle an anticipated increase of passenger traffic due to the changing focus of the war. Modifications included added tables to the transient mess halls and increased baggage-handling capabilities. Esthetic upgrades to the waiting room in the passenger terminal centered on comfort and a more States-like appearance. Furniture, rugs, a newspaper stand, settees and easy chairs made the transition more relaxed.

Declassification of the APO in July 1945 boosted morale. They were no longer "somewhere in India."

In June, Lalmanir Hat claimed the world's longest route of any one twin-engine aircraft station. From Khartoum in Sudan planes and crews flew to Kunming, China, a distance of

4,550 miles. Relief crews from Cairo, Abadan, Karachi, and Lalmanir Hat were enlisted to Lalmanir Hat to assist with the movement of backlogs in equipment and supplies from Agra (India), Barrackpore and Dum Dum (Calcutta) to Khartoum and were assigned to Lalmanir Hat.

A public address system was installed in the passenger terminal to announce incoming/outgoing flights. Information and Education bulletin boards displayed maps of India and various war fronts, plus the latest news items. Radios were installed in the Billeting Office, Mess Hall and Enlisted Men's Day Room. The Mess Hall added new white table cloths and napkins. The installation of a Ladies Powder Room with towels and tissues enriched the Passenger Terminal experience.

In August, came more changes:
- Writing desks and easy chairs were added to the waiting room.
- The PX section carried the basic amenities of soap, shaving cream and toothpaste
- Secure baggage facilities
- Curio shop which was operated by the Jagat Narain and Sons of Calcutta—something to send home as "something from India." 4

Stateside rotations began. In September 1945 the Hope Project started transporting passengers from Chabua and Calcutta to Karachi, with the base mission focused on personnel transport.

11.6 AAFBU 1333rd CHABUA

On 1 Feb 1945, IAW Ops Order #4, 28 Jun, Chabua Wing Headquarters, the passenger shuttle line operating to Assam Wing Bases and Lalmanir Hat began operation from Sookerating instead of Chabua. The facilities at Chabua had become too congested. The shuttle service had actually procured an aircraft with passenger-type seats instead of the web sling-type seats mounted on the inside of the fuselage side walls.

Lt. Bruce L. Magill, Assistant Wing Priorities & Traffic Office (P & T) established a Flight Traffic Clerk training program at Sookerating. Five Flight Traffic clerks were trained to handle Assam Valley Passenger shuttle and Valley Freight. The program allowed the establishment of scheduled flight service to a ± 5-minute accuracy rate, except in cases of weather or maintenance issues. As of Feb. 1, 1945, two separate entities—freight and passenger shuttles—had been combined.[5]

By May 1945 the passenger terminal had a stateside appearance, with wall murals by Capt. Scott, ICDATC artist. Behind the terminal a new transient mess hall was constructed to expedite passenger flow. Chabua had become the busiest base in India.

In 1945 the truck drivers from the first convoy over the Ledo Road were flown back from China and into Chabua. As the war progressed, over 300 convoy drivers were flown back to Chabua every month. [6]

By the end of September, only Chabua and Mohanbari were still flying Hump missions. Those bases not already closed or in the processing of closing had the mission of moving personnel to Calcutta or Karachi.

11.7 AAFBU 1332nd MOHANBARI

With the closure of Sookerating, equipment and truckloads of aircraft spare parts were shipped to Mohanbari. All groups began taking care to pack away tools, equipment, etc. in anticipation of deactivation orders. Squadron D's Mess

Assam Trucking Company

Hall was placed under the supervision of the 1352 AAFBU. Squadron B Mess Hall was closed and the personnel took their meals at Squadron C Mess Hall.

Motor transportation equipment was readied for shipment to Moran. Two mobile laundry units replaced the Quartermaster laundry. Fifty-two British 10-man tents had been taken down in the Squadron B. Area. A survey of base equipment and buildings was made. Several truckloads of timber for the construction of equipment shipping crates were received. The 1116th MP Co. (AVN) transferred to the Northern Air Service Area Command, with their security duties taken over by the security section of Squadron A.

Finally, the base was notified that a definite tonnage allotment for delivery to China had been made. As of 1 October, the allotment tonnage had been cut to 2,000 tons of primarily aviation gas, plus a little subsistence. Only daylight trips would be made and were expected to be completed by 20 October. Deactivation would start as soon as the deliveries were completed. To all there, it was the best news they had had since V-J Day.

The Base Training Office discontinued activities. The Jungle Camp at Kohima, opened under direction of Capt. "Shorty" Grey, received increased interest. The Rest camp at Shillong was closed, creating a waiting list for the Kohima Camp.

When the five Air Freight Terminals were empty and clean, all day-cargo originally destined for China, which had been cancelled, was returned to the shipper. Equipment turn-ins began. A "clean-up truck," used to ensure proper policing of aircraft prior to loading at a dispersal point, was dispatched with five laborers and brooms to clean the cabins and storage spaces at the designated dispersal point.[7]

11.8 AAFBU KUNMING

In late August 1945, ATC provided all transportation from the Kunming staging area for shipment back over the Hump to India until other staging areas opened on the Chinese coast. Personnel leaving China for the U. S. flew back over the Hump to catch a transport ship at Calcutta for home. [8]

Until the fall of Japan, a max effort had been the business of the day. All effort was made to reduce turnaround times and to increase Hump tonnage. With V-J Day came a noticeable change in the operational focus from high-pressure daily trip consciousness to an operational plan of safety and service—the goal was no longer tonnage.

The following restrictions/flight rules were put in place:
- No loaded night take-offs in the C-87, B-24 or the C-109
- Gross takeoff weight for the C-87s and C-109s not to exceed 58,000 pounds.
- Gross takeoff weight for the C-54 day/night operations range from 65 to 68,000 pounds
- Restrictions same for the C-46 as for the C-54, with weights at 45 to 47,000 pounds, dependent on time of take-off and area of operation.
- Instrument flying mandatory for ALL Hump flights
- Fifteen-minute interval separating all aircraft departing any base
- Strict compliance with T.O.s and maintenance directives

With the increase in personnel turnovers, more experienced pilots were sent to other wings. The number of inexperienced replacements didn't cover the number of personnel transferring out of the war zone.

Assam Trucking Company

Rotation points dropped to 70 by October 1. Those with two years in theater and less than 70 points were also shipped home in October. Men started returning to Chabua from China in early September—as many as 400 per day. The billeting at Chabua was inadequate for the number being processed. China personnel were sent to Barrackapore, then on to Karachi. The Wing Headquarters was closed on 15 October 1945.

Operations now centered on the movement of personnel from China to India. Allotment of supplies was curtailed. [9]

Homeward movement began in earnest.

On November 5, 1945, 1/Lt. John Foster sent the long-awaited telegram to his wife. "Home soon. Meet you at Memphis. Stay home until you receive next wire."

Foster actually arrived home in the middle of December, 1945.

Chapter 12.0 — What Happened To ATC?

12.1 FEASIBILITY STUDIES ON THE FLY

The question in the beginning was "can it be done?" Given no real choices or blueprints, the Chinese, in concert with the United States and Great Britain, went about building a theater of war. The history of CBI and ATC illustrates how a non-existent theater and command was finally truly operational by the end of hostilities.

When the aerial resupply of China was undertaken in 1942, not even the most prophetic of men who flew the Hump fully envisioned what military strategic and tactical airlift concepts were being birthed. Most of those involved were engrossed with mission accomplishment and simple survival. Fortunately, many did survive, and of these a few were privileged to continue in military operational, staff and command assignments where they were able to refine the lessons learned on the Hump.

Assam Trucking Company

The Hump effort went from primitive barnstorming to a large-scale undertaking. A grand total of 650,000 tons of supplies and personnel traveled the route over the Himalayas by air. More than half that total was flown to China in the first nine months of 1945. Born of an emergency, it remained an emergency communication system throughout the theater's operations. [1]

In some estimations, while the immediate effect of the ATC's role in the defeat of Japan was deemed questionable, the Theater became the launch pad for several military and civilian capabilities in use today. Without that proven airlift capability, the Berlin Airlift would have been more difficult. General Tunner, who as a Major in 1942 had been involved in the planning effort for the world-wide mission of Air Transport Command, took what he had learned and honed during the Hump operations and applied them to the Berlin Airlift. Furthermore, his experience in the foundation of ATC and in his innovations in Hump Operations translated into the expansion of ATC's role in the USAF and the refinement of airlift capability.

Despite the accomplishments of the ATC in CBI, General George C. Marshall criticized the increase in flights and the need for planes in CBI because of the perceived direct detrimental effect the increase had on the movement of troops and supplies in the European Theater, lengthening the war there. General Marshall felt it directly affected Lieutenant General George S. Patton's armored advance across France to Germany in the late summer of 1944 by restricting the supply of gasoline and munitions. Because of the need for aircraft in CBI, other theaters of the war were assigned only minimal aircraft numbers for the transport of provisions. Many felt that the "real" war was in Europe and, therefore, by extension, the need for war matèriel was more critical for that particular part of the war effort.

On September 18, 1947, the U.S. Air Force was born as a separate United States military organization. In June 1948 the Air Force Command realized the need to find a home for several service-oriented functions. There was reluctance from members of the Air Staff to allow the airlift forces major command status. ATC was redesignated as the Military Air Transport Services (MATS) and was given the responsibilities of airlift/ferrying operations, air communications, air weather service, aeromedical evacuation, photomapping and audio-visual services, and air-sea rescue and recovery worldwide. Two other units, Combat Cargo and Troop Carrier were reorganized under Tactical Airlift Command.

The reorganization was barely accomplished when Russia imposed the land blockade of West Berlin, isolating 3,000,000 freedom-oriented Germans as effectively as Japan's severance of the Burma Road to China in early 1942. Again, Airlift provided the answer under the guidance of General William Tunner, the last commander of the Hump airlift operations. The Soviets administered the worst blow ever dispensed in the Cold War. Military experts credit the Berlin Airlift with stalling Russian expansionist goals for literally "years." In the eighteen months of the airlift undertaking, 2,343,315 *tons* of food, coal, medical supplies, and priority personnel were delivered to the beleaguered city. General Tunner simply applied the lessons learned the hard way in CBI at a time when every resource was made available to him. It is noteworthy that much of Tunner's staff was ATC/Hump trained.

Since that point, the USAF airlift forces have distinguished themselves through hundreds of humanitarian, medical, rescue/recovery, scientific, and military operations in Korea, Viet Nam, Israel, the Middle East, Afghanistan and other political brush fires ignited by those wishing to change the world order.

While keeping pace with the modernization of its fleet, MATS established its own Airlift University at Altus AFB, OK, which today serves as the official airlift schoolhouse. In early WWII, the War Department established in ATC the operation and maintenance of their highest priority airlift unit, the Special Air Mission Wing at Washington National Airport (Ronald Reagan Washington National Airport) and later Joint Base Andrews. The unit's charges include the presidential aircraft known as AIR FORCE ONE.

Tactical airlift capabilities were returned to the fold when the USAF re-designated MATS as Military Airlift Command (MAC), and at the same time assigned it the responsibility of all military airlift. The active fleet, with over 1,000 assets, is augmented by 400 National Guard/Air Force Reserve aircraft and 300 of the most modern jets from the Civil Reserve Air Fleet (CRAF). Active personnel rolls now number 69,200 in the USAF reserves and 106,700 in the Air National Guard. In times of increased airlift requirements, CRAF bring personnel on board to operate and maintain their equipment.

The long-range effects are more apparent. CBI with its "first ever air lift" was the "birthplace and proving ground of mass strategic and tactical air lift capability." The global effectiveness of the United States today is based on the accomplishments of the airmen and transports of ATC in CBI. The greatly-refined rapid deployment units of the U.S. military have their roots in the Air Transport Command, India-China Division of the China-Burma-India Theater of World War II.

Two services, taken for granted now, were developed or honed in CBI. In its beginning Air Search and Rescue was impromptu. As noted earlier, in July, 1943, Capt. John L.

"Blackie" Porter, based at Chabua, India, was assigned two C-47s to conduct official search and rescue missions. The unit became known as "Blackie's Gang." Their mission was to locate downed aircraft, report crash sites and, if possible, render aid in the form of air drop of medical and food supplies, and to guide ground rescue teams to the downed crews.

The need for medical evacuation was the impetus for another service developed for CBI units. Most bases in India had only a dispensary. With the terrain and the lack of ground transportation, the only means of moving the sick and injured was by air. Medical evacuation flights were the rule instead of the exception. A doctor and one or two Army nurses were assigned to such flights to render such emergency medical services as were required during flight.

Airdrop capability in use today was developed to supply ground forces of GALAHAD better known as Merrill's Marauders in Burma. The dense jungle and lack of roads in the area made an airdrop the only means of distribution.

Air Transport Command was the beginning platform for the Military Air Transport Service (MATS) (1948), which became in 1966 Mobility Airlift Command (MAC). Today it is known as the Air Mobility Command (AMC) (1992), with command headquarters at Scott AFB, IL. What was born in the China-Burma-India Theater of World War II, on the fly as little more than an experiment with little time to truly plan, has become one of the major commands of the United States Air Force.

12.2 MILITARY AIRLIFT

12.2.1 Military Air Transport Services (MATS), 1948-1966

- Operation Magic Carpet—Airlift of Moslems stranded in North Africa en route to Mecca (1947) for the Hadj
- Berlin Airlift (1948-49)
- Korean War (1950-53)
- Suez, Lebanon, and Taiwan Straits Crisis (1956-58)
- Operation Deep Freeze (1957-63), codename for a series of United States missions to Antarctica, beginning with "Operation Deep Freeze I" in 1955–56
- Operation New Tape (1960-63)—Congo Airlift
- Big Slam/Puerto Pine (March 1960)—MATS almost doubled its aircraft flying rate during March in a test of its ability to surge to a wartime pace. As part of this exercise, it joined with the Army in the largest peacetime airlift exercise in military history (Big Slam Puerto Pine), airlifting 21,095 troops and 10,925 tons of their combat equipment from the U.S. to Puerto Rico and back.
- Berlin Crisis (1961)
- Cuban Missile Crisis (1962)
- Operation "Big Lift" (1963)
- Vietnam War
- Eagle Thrust
- Nickle Grass (1973)—Airlift of military supplies to Israel during 7-day war

12.2.2 Military Airlift Command (MAC), 1966-1992

- Cold War Operations
  - Safe Haven (1956)—Hungarian refugee airlift
- Vietnam War Operations
  - Homecoming

- Babylift
- New Life (1975)—Evacuation of South Vietnam American supporters from Saigon
- New Arrival (1975)—Second phase of New Life
- Eagle Pull (1975)—Evacuation of Cambodian American supporters from Phnom Penh
* Urgent Fury (1983)—Grenada
* Gulf War
  - Operation Desert Shield (1990)—First need to use CRAF

12.2.3 Air Mobility Command (AMC), 1992 to Present

* Acquisition of Tanker assets from Strategic Air Commands (SAC)
* Second Gulf War
* Afghanistan
* Search and Rescue services morphed into
* Air Rescue Service (ARS)
* Aerospace Rescue and Recovery Service (ARRS) in Vietnam
* Civilian applications. [2]
* Medevac capability in CBI evolved into Aeromedical Transport Wing, with civilian applications in Air Ambulance services across the United States

What was born over the Himalayas, 76 years ago is with us today. The men and women of that underrated and misunderstood theater were the architects and builders of a major military capability today. Many of the children and grandchildren of those first airlifters/Humpsters have been active in various aspects of airlift. The story of the first airlift is over, but the effects goes on.

RETROSPECTIVE

Climbing out of India and flying east toward China three great river gorges—the Salween, the Irrawady, and the Mekong—and the mountains, with peaks from 15,000 to 20,000 feet, of the main Hump and the Santung Range, were crossed daily from 1942-1945 by the men of the Air Transport Command. Only thick green jungles were to be seen below with no signs of life. After passing the Mekong River, signs of civilization could again be observed from the air.

These same areas would become familiar to another generation who fought in Vietnam. Many of the men and women who fought there were either veterans of the Hump or the children of "Humpsters" (as they would call themselves). The comparisons are not lost.

While reading through my father's military records which had been saved by my mother in her small suitcase she carried from camp-to-camp while my father was in aviation cadets, I began to notice how much of my life was shaped by that long-ago effort.

In March of 1959, at Castle AFB, then the Strategic Air Command (SAC) training center, my father, a KC-97 pilot, retrained as a KC-135 pilot. It was there that I "flew" my first ever simulator—an office chair in front of a pilot console. It was a prophetic flight.

A few months later in August, at the age of 12, our family boarded a MATS (Military Air Transport Service formerly ATC) Constellation aircraft for the five-hour flight to Puerto Rico. While at Ramey AFB, we experienced the Big Slam Puerto Pine exercise in 1960, and in August 1962 rotated to the States again on a MATS Constellation. In 1966 MATS became the Military Airlift Command (MAC).

My father retired from the USAF in 1970 with 28-years of service. His last duty assignment was at Altus AFB, OK, which by then had become the school house for MAC. Altus was almost a second home to my family. My youngest brother, a KC-135 pilot himself, recommended me for a job with a government contractor in charge of C-5 training at Altus. Now the training center for the C-5, Galaxy, and later airlift platforms, i.e., C-141 Starlifter, C-17, Globemaster III, the KC-135 Stratotanker and soon to join the arsenal, the KC-46 Pegasus tanker, Altus AFB had been chosen for the airlift training mission because of the number of days of clear air, view unlimited (CAVU) conditions.

As an English teacher and a training designer and developer for the U. S. Army, I took the job at Altus AFB, OK, in 1993 as a Courseware Developer, writing and editing training materials for pilots, copilots, flight engineers, and loadmasters of the largest plane in the U. S. arsenal—the C-5. In 1994 I moved to another contract, again designing and developing training materials for pilots, copilots, navigators and boom operators for the KC-135. This time the simulator was on a motion base and stood three stories high. No office chair was in place. It had been 45 years since my first sim flight and the equipment had grown in size and scope. Later I went to yet another contract to develop training for the C-17.

Assam Trucking Company

After my mother died and I was given the treasured 64-year old suitcase, I was surprised to find copies of the training materials used by my father at Castle, his training records for the KC-135, and flying charts for Puerto Rico. I had come full circle.

APPENDICES

B. F. Bates

A Pilot's Check List

C-47, C47A, C-53, C-53C, Airplanes R-1830-92 Engines (Navy R4D-1, R4D-3, R4D-5)

BEFORE STARTING ENGINES

1. Check Form I and status today. (Navy Yellow Sheet.)
2. Check Form F, Weight and Balance Clearance, AN 01-1-40.
3. Wheels chocked.
4. Landing gear pins installed.
5. Determine whether carburetor air filters are installed.
6. Cowl flaps "OPEN."
7. Mixture "IDLE CUT-OFF."
8. When fuselage tanks are installed, valves should be "CLOSED."
9. L.H. fuel selector to L.H. main or to fullest tank.

    R.H. fuel selector to R.H. main or to fullest tank.
10. Check fuel quantity gages.
11. Cross feed "OFF."
12. Propeller controls "LOW PITCH."
13. Carburetor heat controls "COLD."
14. Oil shutter controls as required.

15. Battery cart "ON."

16. Instrument switch "ON."

17. Wobble up fuel pressure.

18. Prime 2 to 3 seconds with outside temperature 40° to 60° F (4.4° to 15.6° C).

    Below 40° F (4.4° C), 5 to 8 seconds. With primers in nacelle, manual priming necessary.

19. Ignitions switches "ON."

20. Energize, "ENGAGE STARTERS."

DURING WARM-UP

1. When engine catches, mixture control "AUTO RICH."

2. Warm up at 800 to 1000 rpm.

3. Check engine instruments in green.

4. Master battery "ON."

5. Move propeller pitch controls to "FULL HIGH PITCH" at least twice with manifold pressure at least 25 inches Hg rpm should drop to 1200 rpm.

BEFORE TAKE-OFF

1. Remove landing gear pins after moving landing gear control handle from "NEUTRAL" to "DOWN," then back to "NEUTRAL" to equalize pressure in struts.

2. Set L.H. engine fuel selector to L.H. main or to tank with most fuel.

Set R.H. engine fuel selector to R.H. main or to tank with most fuel.

3. Mixture in "AUTO RICH" ("TAKE-OFF" and "CLIMB" position).

4. Carburetor heat "COLD."

5. Oil cooler shutters closed as required.

6. Propeller pitch control "FULL LOW PITCH."

7. Cross feed "ON."

8. Landing gear latch lever to "SPRING LOCKED" position.

9. Trim tabs set to "ZERO."

10. Tail wheel "LOCKED."

11. Cowl flaps "TRAIL."

12. Tighten throttle lock.

13. Check all Flight Controls for FREE MOVEMENT.

DURING FLIGHT

1. Retract gear-control "NEUTRAL."

2. Cross feed "OFF."

3. Cowl flaps "TRAIL" FOR CLIMB. Closed to "OFF" for cruise.

4. Mixture "AUTO RICH" for take-off and climb.

5. Take off and indicate 110 mph or 120 mph as soon as possible.

6. FLIGHT INSTRUCTION

CONDITION	ENGINE RPM	MANIFOLD PRES. & MIX 100 OCT.	PRES & MIX 91 OCT.	MAX. CYL HEAD TEMP
Take-off	2700	48" Hg Auto Rich	42.7" Hg Auto Rich	260° C
Normal Rated Power	2550	40" Hg Auto Rich	38.6" Hg Auto Rich	260° C
Maximum Cruising	2230	33" Hg Auto Lean	27.5" Hg Auto Rich	232° C
Desired Cruising	2150	28" Hg Auto Lean	28" Hg Auto Rich	232° C
Cruise for Minimum Specific Fuel Flow	1700	30" Hg Auto Lean	30" Hg Auto Lean	232° C

BEFORE LANDING

1. L.H. fuel selectors to L.H. main or fullest tank.

    R.H. fuel selectors to R.H. main or fullest tank.

2. Carburetor heat "COLD."

3. Mixture "AUTO RICH" for take-off and—climb.

4. Gear "DOWN," control "NEUTRAL," and "LATCHED," green light.

5. Cross feed "ON."

6. Cowl flaps "CLOSED" and in "OFF" position.

7. Set propeller controls to 2230 rpm.

8. Extend flaps (max. speed for extension 112 mph). Use full flap unless gusty condition on cross wind.

AFTER LANDING

1. Propeller "LOW PITCH."

2. Flaps "UP."

3. Tail wheel "UNLOCKED" for turning and taxiing.

4. Cowl flaps "OPEN."

5. Use oil dilution as required. T.O. No. 02-1-29.

6. Mixture control "IDLE CUT-OFF" with engine turning 1000 rpm.

7. Ignition "OFF."

8. Fuel selector "OFF>'

9. Electrical switches "OFF."

10. Landing gear lever "DOWN."

11. Cowl flaps "CLOSED" when engine cools.

12. Set PARKING BRAKES when cooled—install landing gear pins.

13. Wheels chocked—surface controls installed.

Revised September, 10, 1943.

Supersedes Pilot's Check Lists of Previous dates.

B GLOSSARY

Acronyms

AA	Anti-Aircraft guns
AAB	Army Air Base
AAC	Army Air Corps
AAF	Army Air Force
AAFBU	Army Air Force Base Unit
AAF-IBT	Army Air Force, India-Burma Theater
AC	Aviation Cadet
ACS	Air Corps Supply
AFB	Air Force Base
A/E	Aerial Engineer
AMC	Air Mobility Command
ANG	Air National Guard
AO	Area Orderly
APO	Army Post Office
APOC	Air Part Out-of-Commission
ASC	Air Service Command
ATC	Air Transport Command
AvCad	Aviation Cadet
AVG	American Volunteer Group, "Flying Tigers"

AVN	Aviation
AWOL	Absent Without Official Leave
BG	Brigadier General
Bn	Battalion
Capt.	Captain
CAVU	Ceiling and Visibility Unlimited
CBI	China-Burma-India
CC	Combat Cargo
CG	Center of Gravity
CNAC	China National Aviation Corporation
CO	Commanding Officer
CQ	Charge of Quarters
CRAF	Civilian Reserve Air Fleet
CTD	College Training Detachments
CWO	Chief Warrant Officer
DDT	Synthetic chemical compound used as a pesticide
DFC	Distinguished Service Cross
DNIF	Duties Not Including/Involving Flying

DR	Dead Reckoning
Dzus	Quick turn fastener
FDR	Franklin Delano Roosevelt, President
F/O	Flight Officer
G.I.	Government Issue
IAW	In Accordance With
ICD	India-China Division
ICD-ATC	India-China Division, Air Transport Command
ICW	India-China Wing
ICW-ATC	India-China Wing, Air Transport Command
I&E	Information and Education
I&S	Intelligence and Security
IFF	Identification Friend or Foe
KMT	Kuomintang
LORAN	Long Range Navigation
MAC	Military Airlift Command
MAE	Medical Air Evacuation

MATS	Military Air Transport Service
MIA	Missing in Action
MOS	Military Occupational Specialty
MP	Military Police
MPH	Miles Per Hour
MX	Maintenance
NOTAM	Notice to Airmen
O_2	Oxygen
OBE	Overcome-By-Events
OCS	Officer Candidate School
OLC	Oak Leaf Cluster
Ops	Operations
OTU	Operational Training Unit
Pan Am	Pan American Airlines
Pfc	Private First Class
PLM	Production Line Maintenance
PSI	Pressure/Square Inch
P&T	Priorities and Traffic
Pvt	Private

PX	Post Exchange
QM	Quartermaster
RAF	Royal Air Force
RN	Registered Nurse
R/O	Radio Operator
ROTC	Reserve Officer Training Corps
R&R	Rest and Recreation
SAM	Special Air Mobility Wing (Air Force One)
SOP	Standard Operational Procedure
TBA	To Be Announced
TO	Technical Order (maintenance manual)
TO&E	Table of Organization and Equipment
USAAC	United States Army Air Corp
USAAF	United States Army Air Force
USAF	United States Air Force

USO	United Service Organizations—support organization for military service members and military families
VD	Venereal Disease
VE	Victory in Europe
VJ	Victory in Japan
WX	Weather

DEFINITIONS

Aluminum Trail	Route nickname
Hump	Route nickname
Rockpile	Route nickname
Bearer	Porter
Bren Gun	Anti-aircraft gun
Chowkidar	Watchman, gatekeeper
C-Rations	Prepared and canned wet food ration
Dhobi	Indian caste, whose primary job was to do laundry.

Dhobi itch	Skin rash generally in the groin area related to the betel nut chewed by the Dhobi laundry crew
Dragon	Bad luck following the Chinese
Flying Tigers	Also knowns as the AVG American Volunteer Group, headed by General Claire Chennault. A mercenary group.
Indian Pioneers	Indian soldiers trained and equipped for road, rail and engineering work as well as conventional infantry service.
Jim Crow	Facilities and programs provided were separate but equal, supporting racial segregation.
Kachin	Natives of northwestern Burma (Myanmar) from Myitkyina district
Lister Bag	Canvas container, used especially for supplying troops in the field with pure water.
Lend-Lease	Principal means for providing U.S. military aid to foreign nations during World War II.

Max Effort	Maximum airlift effort to maintain flight movements
Monsoon	Violent weather pattern characterized by turbulent winds and torrential rainfall
NOTAM	Notice filed with an aviation authority to alert aircraft pilots of potential hazards along a flight route or at a location that could affect the safety of the flight
Parapack	A parachute backpack.
Pucca	Building designed to be permanent, unlike bashas
Revetments	Parking area for one or more aircraft, surrounded by blast walls on three sides.
Slit Trench	A narrow, rather shallow trench for protecting a soldier from shellfire, bombs, etc.
Tug	Powered equipment to move aircraft on ramps or in tight hangar space.
Walla(h)	Indian laborer, trained
Weasel	All terrain track vehicle
Wog	Indian native laborer

C END NOTES

Chapter 1—Prologue

[1] Tunner, William H., Over the Hump, p. 43

[2] "Air Transport Command," *The Army Air Forces in World War II*, p. 14.

[3] Schultz, Duane P., The Maverick War. New York, New York: St. Martins Press, 1987, p. 56.

[4] Schultz, Duane P., The Maverick War. New York, New York: St. Martins Press, 1987, p. 56.

[5] Ibid., 54-57; and,

Schultz, p.13.

[6] Boxer Protocol was signed in 1901 to end the Boxer Rebellion and reassert a foreign government's right to have soldiers in China to protect the diplomatic mission of that country., Toland, John, The Rising Sun, p.46.

[7] Madam Chiang and her brother, T. V. Soong, were from a very prominent family led by Charlie Soong, family patriarch, American-educated Methodist minister and his wife, Lady Xu, a descendant of the Ming Dynasty. The Soong siblings had been educated in the United States and understood the U.S. culture and politics very well. Soong Dynasty, p. 362

Three Soong sisters married well. Ai-Ling married H. H. Kung, China's finance minister; Mei-Ling married Chiang Kai-shek; and, Ching-Ling married Sun Yat-sen. "One Loved money, one loved power and one loved China." Famous Chinese saying about the three sisters. Schultz, The Maverick War, pp. 22-23.

[8] Seagrave, Sterling, The Soong Dynasty. New York, New York: Harper and Row, 1985, p. 362

[9] Seagrave, Sterling, The Soong Dynasty. New York, New York: Harper and Row, 1985, p. 359.

[10] Seagrave, p. 361.

[11] Schultz, The Maverick War, p. 3

[12] Air Operations in India and Burma, Craven and Cates-1, p. 1

[13] Ibid

[14] Air Operations in India and Burma

[15] Seagrave, Sterling, The Soong Dynasty.

[16] Futrell, Development of Aeromedical Evacuation, p. 77

[17] Sherwood, Roosevelt and Hopkins: Air Intimate History, p. 513

[18] Air Ops, India-Burma

[19] Ibid

[20] Army Military World War II, Chapter 23, "Other Overseas Theaters," pp. 464-465

[21] Army Military World War II, Chapter 23, "Other Overseas Theaters," pp. 464-465

[22] Stilwell, Joseph, The Stilwell Papers. New York: William Sloane Associates, Inc., 1948. P. 118 Sherwood, p. 513.

23 Army Air Force at War, p. 88.

24 Army Air Force at War, p. 89.

25 White, Edwin Lee, Ten Thousand Tons by Christmas, p. 59.

26 Foster, John M. Lt. Col USAF (ret).

27 Thorne, Bliss K.

28 Seagrave, Sterling, The Soong Dynasty

29 Ibid

30 Stilwell, Joseph, The Stilwell Papers

31 Air Ops, India-Burma

32 Ibid

33 Ibid

Chapter 2—The Gauntlet

1 Heck, Airline to China, p. 116

2 Ibid., p. 116

3 1333rd Unit History, Misamari

4 1332nd Unit History, Mohanbari

5 1333rd Unit History, Misamari

6 China Hump Pilot's Association, Vol. I, p. 63, 184

7 Heck, p. 120

8 Ibid, p. 123

9   China Hump Pilot's Association, Vol. I, p. 354

10 Tunner, William H., Over the Hump, p. 58

11 China Hump Pilot's Association, Vol. II, p. 354

12 Tunner, p. 47

13 Ibid. p.43

14 Johnson, Lonnie, Interview

15 Army Air Forces in World War II, Weather, p. 201

16 "Let-down" was the term then for the term "on approach" used today.

17 1333rd Unit History

18 In June 1942, the American Volunteer Group (AVG) was incorporated into the U.S. Army Air Corps as the 14th AAF but was still known as the Flying Tigers. Chennault remained in command of the group but as a U.S. AAF Colonel was later promoted to Brigadier General.

19 Foster, interview

20 Thorn, Bliss K, The Hump, pp. 49

21 Ibid., p. 31

22 Interview, Foster

23 Foster; Ex-CBI Roundup Jan '83, "Dragon", p. 10

24 New York Times, April 21, 1945, p. 8

25 Interview, Leonard

26 Jonasson, AAF Weather Service, p. 325

27 Thorne, p. 120

28 Thorne, Bliss K., p.109.

29 Tunner, p. 75

30 Interview, Leonard

31 1333rd Unit History, Misamari

32 China Hump Pilot's Association, Vol. I, p. 184

33 1333rd Unit History, Misamari

34 Army Air Forces in World War II, Weather, p. 125

35 1333rd Unit History, Misamari

36 Ibid.

37 Ibid.

38 Thorne, p. 115

39 Jonnason, p. 325

40 Tunner, p. 75

41 Thorne, p. 49-50

42 Heck, p. 239

43 Interview, Foster

44 Army Air Forces in World War II, Weather, p. 120

45 1333rd Unit History, Misamari

46 Interview, Leonard

47 China Hump Pilot's Association, Vol. II, p. 189

48 1333rd Unit History, Misamari

49 Ibid.

50 1332nd Unit History, Mohanbari

51 1333rd Unit History, Misamari

52 1332nd Unit History, Mohanbari

53 1333rd Unit History, Misamari

54 Ibid.

55 Foster Interview

56 1328th AAFBU communication sent on 26 November 1943 concerned fighter protection.

Headquarters,

Eastern Sector

INDIA CHINA WING AIR TRANSPORT COMMAND

Subject: Fighter Protection for ATC Aircraft

TO: Commanding Officer, HQ Project #8 ATC, Misamari, India.

1. Information had been sent to all groups that maximum fighters protection is being forwarded north of the Sumprabum area for all ATC aircraft. This protection has been functioning in a very satisfactory manner, however, our ATC pilots are, in many cases, seriously abusing and thwarting the efforts of the protective forces being provided, by violating prescribed flight routes between Assam and China. Unmistakable directives have been sent out requiring transports to stay well north of Sumprabum. Reports continue to be received that transports are flying south of Sumparbum and south of fighter patrols. An intelligence report received this date is quoted: "Fighters on patrol, far to the north on sighting these planes are compelled to regard them as enemy planes and so fly south to

Assam Trucking Company

destroy them. On getting close to the transport planes, they recognize them as friendly. In some cases they even get the numbers of the planes, which are transmitted to headquarters. The fighters feel very keenly about this situation, as these unexpected and unnecessary diversions interfere with their patrol schedules and consume so much gasoline that many fighters get back to base with only 5 minutes supply of fuel."

2. Since there is no normal reason for ATC aircraft to get as far south as Sumprabum, severe disciplinary action will be taken in such cases reported to this headquarters. Operations Officers will personally see that these instructions are thoroughly understood by all pilots.

By order of Colonel Hardin

/t/ John R. Kilgore

Lt. Colonel, Air Corps

Executive

[57] Heck, p. 117

[58] White, Edwin, Ten Thousand Tons by Christmas, p 71

[59] Army Air Forces in World War II, Weather, p. 109

[60] China Hump Pilot's Association, Vol. I, p. 63

[61] Air University, AF ROTC History of the USAF, p. 6

[62] Ibid.

[63] Tunner, William H., Over the Hump, p. 51

[64] Numbers for ATC AAFBUs were compiled from records in China Airlift—The Hump, Volume I, pp.

619-622. These numbers do not include the losses of any other units flying the Hump route between 1943 and 1945. No direct combat losses were noted.

65 Brig. Gen. Tunner would go on to become the commander of Berlin Airlift 1948-49 and the Military Air Transport Service (MATS). MATS was the successor to ATC and the predecessor of the Military Airlift Command (MAC) and the Air Mobility Command (AMC) of today.

66 China Hump Pilot's Association, Vol. I, p. 63.

67 On 13 July 1945, Headquarters India Burma Air Service Command announced a reduction of the number of hours required for the Air Medal and the DFC. The new criteria were now:197

Medal	Hours	or	Flights
Air Medal	**50**		**25**
1st OLC to AM	100		50
2nd OLC to AM	250		100
3rd OLC to AM	350		150
4th OLC to AM	450		200
5th OLC to AM	550		250
DFC	**200**		**75**
1st OLC to DFC	300		125
2nd OLC to DFC	400		175
3rd OLC to DFC	500		225

Chapter 3—Training

1. Army Recruiting Induction Station, Tulsa, OK. September 28, 1942. Forms required and Instructions for making application for Aviation Cadet program.

2. Army Air Force, "Air Transport Command, Finding the Pilots" pp. 29-30

3. McCoy, William, Interview.

4. Army Air Force, "Air Transport Command," pp. 36-38.

5. Army Recruiting Induction Station, Tulsa, OK. September 28, 1942.

6. Aviation Cadet Program, 1940-47

7. USAAF AvCad Program

8. Hunt, Jack and Fahringer, Ray, Student Pilot Handbook, 1943, p. 5.

9. Public Law 658

10. Army Air Force WWII, p. 42

11. Ibid.

12. Ibid.

13. AAFBU 1329th, Misamari Unit History

14. "Vacationland in the Jungles of the Himalayas," Declassified Training Syllabus, Jungle Indoctrination Camp.

15. 1333rd AAFBU, Chabua Unit History

16 Foster, Interview
17 1326th AAFBU, Lalmanir Hat Unit History
18 1328th AAFBU, Tezpur Unit History
19 Unskilled local laborer, i.e., Chinese or Indian
20 1333rd AAFBU, Chabua Unit History

Chapter 4—AAFBUs

1 Development of the Aerial Ferry, Unclassified.
2 Thorne, Bliss K. The Hump. New York: J. B. Lippincott Company, 1965 p. 23
3 1333rd AAFBU, Chabua Unit History
4 Thorne, p. 24
5 1333rd AAFBU, Chabua Unit History
6 1332nd AAFBU, Mohanbari Unit History
7 1333rd AAFBU, Chabua Unit History
8 1328th AAFBU, Misamari Unit History
9 China Hump Pilot's Association, Vol I, p. 217
10 1328th AAFBU, Misamari Unit History
11 1329th AAFBU, Deragon Unit History
12 1333rd AAFBU, Chabua Unit History
13 Leonard Interview
14 Foster Photo

15. Tunner, William, Over the Hump, pg. 55
16. Foster, interview
17. Leonard, Gordon J, Interview October 12, 1986.
18. Foster
19. 1328th AAFBU, Misamari Unit History
20. HDQT. 1332nd AAFBU ATC 23 Sept 1944 Malaria Discipline
21. 1330th AAFBU, Jorhat Unit History
22. 1326th AAFBU, Lalmanir Hat Unit History
23. 1333rd AAFBU, Chabua Unit History
24. 1328th AAFBU, Misamari Unit History

Chapter 5—Morale

1. Aerial Ferry, Declasssified, pp. 38-40-
2. Tunner, William H., Lt. General, Over the Hump, p.
3. AAFBU 1328th, Misamari Unit History 48
4. AAFBU 1333rd, Chabua Unit History 43-49 pp 45-46
5. Interview, Paul Schaffer
6. Interview, Gordon J. Leonard
7. AAFBU 1333rd, Chabua note card 37
8. AAFBU 1326th, Lalmanir Hat Unit History 14 and 22, 36, 26,

9   AAFBU 1330th, Jorhat Unit History nc. 14

10 AAFBU 1330th, Jorhat Unit History, pp. 8-11

11 AAFBU 1328th, Misamari Unit History, pp. 34-35

12 Cullum /CBI Manhunt, Karl W. Weidenburner 2014

13 AAFBU 1326th, Lalmanir Hat Unit History 5

14 AAFBU 1326th, Lalmanir Hat Unit History 6/45

15 AAFBU 1332nd, Mohanbari Unit History

16 Tunner, p. 89

17 AAFBU 1333rd, Chabua Unit History 29-30

18 AAFBU 1333rd, Chabua Unit History 26

19 AAFBU 1330th, Jorhat Unit History, p. 69

20 AAFBU 1328th, Misamari Unit History 75-78 paper

21 AAFBU 1332nd, Mohanbari Unit History

22 AAFBU 1326th, Lalmanir Hat Unit History nc 6

23 AAFBU 1328th, Misamari Unit History, p. 15, 9 5/45

24 AAFBU 1326th, Lalmanir Hat Unit History

25 AAFBU 1332nd, Mohanbari Unit History

Chapter 6—Planes

1   Army Air Force in WWII, "Finding the Planes"

2   Aerial Ferry, Declassified

3 "Finding the Planes," p. 22.

4 Ibid, p. 21.

5 AAFBU 1332nd, Mohanbari, Unit History

6 AAFBU 1333rd, Chabua Unit History

7 AAFBU 1332nd, Mohanbari Unit History

8 AAFBU 1333rd, Chabua Unit History

9 Gordon Smith, Interview

10 China Hump Pilots, Vol. II, p. 189.

11 Interview, Foster

12 AAFBU 1333rd, Chabua Unit History

13 Interview, Gordon Smith

14 China Hump Pilots, Vol. II, Herb Fisher, p. 191

15 China Hump Pilots, Vol. II, p. 354.

16 Heck, p. 124.

17 Ibid., p. 136.

18 "Finding the Planes", pp. 27-28.

19 China Hump Pilots, Vol. I, p 155.

20 AAFBU 1333rd, Chabua Unit History

Chapter 7—Maintenance

1 AAFBU 1333rd, Chabua Unit History

2 Gordon Leonard, Interview

3   Interviews, Gordon Smith and John M. Foster

4   AAFBU 1326th, Jorhat Unit History

5   AAFBU 1328th, Misamari Unit History

6   Leonard Interview

7   AAFBU 1333rd, Chabua Unit History

8   AAFBU 1328th, Misamari Unit History

9   Hump Express, November 1945, Official Newspaper of the India-China Division, Air Transport Command, APO NY

10   AAFBU 1328th, Misamari Unit History

11   AAFBU 1330th, Jorhat Unit History

12   AAFBU 1326th, Lalmanir Hat Unit History

13   HPA Newsletter Autumn 1993, Memories of Ed Crumpacker

14   China Hump Pilots, "Buzzard Bait," Vol. 3, p. 131.

15   1330th AAFBU, Jorhat Unit History

Chapter 8—Logistics

1   AAFBU 1332nd, Mohanbari, Unit History

2   Interview, Foster

3   AAFBU 1332nd, Mohanbari Unit History

4   OHP pp. 102-105

5   AAFBU 1328th, Misamari Unit History

Assam Trucking Company

6   W. McCoy, Interview

7   AAFBU 1328th, Misamari Unit History

8   Army Air Force in World War II, Heck, p. 146

9   AAFBU 1333rd, Chabua Unit History

10   AAFBU 1332nd, Mohanbari Unit History

11   AAFBU 1333rd, Chabua Unit History

12   AAFBU 1332nd, Mohanbari Unit History

13   AAFBU 1305th, Dum Dum Unit History

14   Army Air Force in World War II, Heck, p. 146

15   Thorne, p. 4

16   Ibid., p. 119

17   Ibid., p. 145

18   Ibid., p. 135

19   "Myitkyina," New York Times, May 21, 1944, p. 27.

20   Thorne, p. 119

21   Tunner, p 120

Chapter 9—Operations

1   "Air Supply Record," New York Times, April 8, 1945, p. 8.

2   "U.S. Army Railroad," New York Times, 5/4/1944, p. 10.

3   "Myitkyina," New York Times, 5/21/1944, p. 27.

4   AAFBU 1332ⁿᵈ, Mohanbari Unit History
5   "U.S. Fliers," NY Times 1/21/1945, p. 7.
6   AAFBU 1326ᵗʰ, Lalmanir Hat Unit History
7   AAFBU 1333ʳᵈ, Chabua Unit History
8   Craven & Cate, Combat Chronology, pp. 539-540.
9   AAFBU 1328ᵗʰ, Misamari Unit History
10 "U.S. Fliers," NY Times 1/21/1945, p. 7.
11 AAFBU 1328ᵗʰ, Misamari Unit History

Chapter 10—Search and Rescue

1   AAFBU 1326ᵗʰ, Lalmanir Hat Unit History
2   Ibid.
3   China Hump Pilots Vol. I, p. 119 & II p. 303
4   China Hump Pilots, pp. 188-89
5   AAFBU 1328ᵗʰ, Misamari Unit History
6   AAFBU 1332ⁿᵈ, Mohanbari Unit History
7   AAFBU 1333ʳᵈ, Chabua Unit History
8   AAFBU 1328ᵗʰ, Misamari Unit History
9   China Hump Pilots, Vol. 1, p 238
10 AAFBU 1328ᵗʰ, Misamari Unit History
11 AAFBU 1332ⁿᵈ, Mohanbari Unit History
12 AAFBU 1328ᵗʰ, Misamari Unit History

[13] AAFBU 1330th, Jorhat Unit History

[14] AAFBU 1326th, Lalmanir Hat Unit History

[15] China Hump Pilots, Vol 1, pp. 110-112

[16] Medical Support of the AAF in WWII, p. 876

Chapter 11—Closing Up Shop

[1] AAFBU 1329th, Deragon Unit History

[2] AAFBU 1330th, Jorhat Unit History

[3] AAFBU 1327th, Tezpur Unit History

[4] AAFBU 1328th, Misamari Unit History

[5] AAFBU 1333rd, Chabua Unit History

[6] AAFBU 1327th, Tezpur Unit History

[7] AAFBU 1332nd, Mohanbari Unit History

[8] "Staging Area," NY Times 8/28/1945, p. 3

[9] AAFBU 1333rd, Chabua Unit History

Chapter 12—Where Did ATC Go?

[1] Heck, Frank H. "Airline to China," pp. 150-51.

[2] Military Airlift Command, Office of History, pp. 86, 98-99, 109, 130, 147, 150, 152, 154, 180, 191.

D. Bibliography

WORKS CITED

Craven and Cates. 1958. *The Army Air Forces in World War II.* Edited by Wesley Frank and Cates, James Lee Craven. Vol. 1. 9 vols. Chicago, IL: University of Chicago Press.

Downie, Don. 1983. "The First Real Airlift." *Ex-CBI Roundup*, Jan, 1 ed.

Foster, Thomas Randall, Col. USAF (ret). n.d. *Localizer.*

Futrell, Robert. 1961. "Development of Aeromedical Evacuation." Study, USAF Historical Research Agency.

Harmon, Fred. 1944. *The Cadet.* Bruce Field, Ballinger, Texas: Army Air Forces Training Detachment.

Heck, Frank H. 1958. *The Army Air Force in World War II: Service Around the World.* Edited by Wesley Frank and Cates, James Lee Craven. Vol. 7. 9 vols. Chicago, IL: University of Chicago Press.

n.d. *History of the USAF.* Air Force ROTC, Maxwell AFB: Air University.

Howton, Harry, interview by Barbara Wilson. 1987. *HPA*

Hump Express. 1945. "Assembly Line Aircraft Repair Started at 1330." November 15.

Hump Pilot's Association. 1980. *China-Burma-India Hump Polit's Asscoiation.* Vol. 1. 3 vols. Poplar Bluff: HPA.

Hunt, Jack and Fahringer, Ray. 1943New York. *Student Pilot Handbook.* New York: Books, Inc.

John M. Foster, LTC USAF (ret), interview by Barbara F. Wilson. 1987. *CBI Experiences* (April).

Johnson, Lonnie, interview by Barbara Wilson. 1984. *Misamari 1945*

Jonasson, Jonas. 1958. *The Army Sir Forces in World War II; The AAF Weather Service.* Edited by Wesley Frank & Cate, James Lee Craven. Vol. 7. 9 vols. Chicago, IL: University of Chicago Press.

Leonard, Gordon J., interview by Barbara Wilson. 1987. *Jorhat 1944-45*

Loving, Robert M "Pete". n.d.

McCoy, William, interview by Barbara Wilson. 1987. *Jorhat 1943-44*

New York Times. 1944. "Myitkyina." May 21: 7.

New York Times. 1945. "Staging Area." August 28: 3.

New York Times. 1944. "U.S. Army Railroad." May 4: 10.

New York Times. 1945. "U.S. Fliers." January 21: 7.

Romanus, Charles F. and Sunderland, Riley, ed. 1955. *U.S. Army in World War II.* Vol. 9. 9 vols. Washington, D.C.: Department of the Army.

Romanus, Charles F., and Sunderland, Riley, ed. 1955. *U.S. Army in World War II.* Vol. 9. 9 vols. Washington D.C.: Department of the Army.

Schaffer, Paul, interview by Barbara Wilson. 1985. *Kurmitola 1944-45*

Schultz, Duane P. 1987. *The Maverick War.* New York: St. Martins Press.

Seagrave, Sterling. 1985. *The Soong Dynasty.* New York: Harper and Row. Accessed 1994.

Smith, Gordon, interview by Barbara Wilson. 1984. *Crew Chief*

Stilwell, Joseph, General USA (ret.). 1948. *The Stilwell Papers.* New York: William Sloane Associates.

Thorne, Bliss K. 1965. *The Hump: The Great Military Lift of Worl War II.* Philadelphia: J. B. Lippincott Company.

Toland, John. 1970. *The Rising Sun: The Decline and Fall of the Japanese Empire 1936-45.* New York: Random House.

Tunner, William M., Maj Gen, USAF, (ret). 1964. "Over the Hump." New York: Duell, Sloan, and PPearce.

U.S Army Air Forces, Air Transport Command. 1945. "AAFBU 1327th Unit History." Tezpur, Assam: Air Transport Command, Jan-Aug 1. 1-13. Accessed October 4, 1994.

U.S. Army Air Force, Air Transport Command. 1944. "AAFBU 1328th Unit History." Unit History, Command Historian, Air Mobility Command,

Assam Trucking Company

Misamari, Assam, India, 2-6. Accessed October 6, 1994.

U.S. Army Air Force, Air Transport Command. 1945. *AAFBU 1328th, Unit History.* Unit History, Scott AFB: Air Mobility Command. Accessed October 6, 1994.

U.S. Army Air Force, Air Transport Command. 1944-45. *AAFBU 1329th, Unit History.* Unit History, Air Mobility Command, Command Historian Office, Scott AFB: Air Mobility Command. Accessed October 7, 1994.

U.S. Army Air Force, Air Transport Command. 1944-1945. *AAFBU 1330th, Unit History.* Unit History, Air Mobility Command, Command Historian Office, Scott AFB: Air Mobility Command. Accessed October 5, 1994.

U.S. Army Air Force, Air Transport Command. 1942-1943. *AAFBU 1332nd, Unit History.* Unit History, Command Historian, Scott AFB: Air Mobility Command. Accessed October 4, 1994.

U.S. Army Air Forces, Air Transport Command. 1943. "AAFBU 1328th Unit History." Edited by AMC Historian Office. Misamari, Assam: Air Mobilty Command, September-December 1. 1-24. Accessed October 6, 1994.

U.S. Army Air Forces, Air Transport Command. 1945. *AAFBU 1332nd, Unit History.* Unit History, Air Mobility Command, Command Historian, Scott AFB: Air Mobility Command. Accessed October 4, 1994.

U.S. Army Airc Forces, Air Transport Command. 1942-1945. "AAFBU 1326th Unit History." Lalmanir Hat, Assam: Air Mobility Command, May-Sept 1. 1-21. Accessed October 4, 1994.

U.S. Coast and Geodetic Survey. 1944. "AAA Flight Chart, Kun Ming, Yunnan, China to Chabua, Assam, India, No. 133." *Aeronautical Chart Services.* Washington, D.C.: Headquarters, Army Air Forces, January.

U.S.Army Air Force, Air Transport Command. 1942-1945. *AAFBU 1333rd, Uit History.* Unit History, Air Mobility Command, Command Historian Office, Scott AFB: Air Mobility Command. Accessed October 5, 1994.

1943. "Vacationland in the Jungles of the Himalayas." *Syllabus.*

W.D.A.G.O. 1942. "Soldier's Individual Pay Record." *Form No. 28, Allotment.* March 26.

White, Edwin Lee. 1970. *Ten Thousand Tons by Christmas.* Burlington, VT: Vantage Press.

INDEX

14th Air Force [American Volunteer Group)AVG), Flying Tigers,

    22, 28, 37, 45, 53, 162, 167, 176, 203

16th AFR Station VU2ZK, 106

p 244-45102

24th Airways Det., 86

90-day wonders, 66

Abors, 48

ACTIVITIES, 78, 104-05, 107, 109, 118, 143 *See*

advanced twin-engine training, 71

Agnew, Capt. Sanford, 116

Agra, 148, 153 205

Air Medal, 62, 64-65, 67, 108-09, 197

Air Service Command, 61, 124

Air Transport Command, 21,

    35-36, 61-62, 68, 72, 94, 97, 108-09, 122, 124, 137, 154, 159, 163, 168, 181, 215, 217-18, 221

Air Transport Command, 159, 163, 168, 181, 215, 217-18, 221

Alexander, Brig. Gen./Col E. H., 63, 74, 98

Allison, Charles G., 180

allotment, 69, 207

Almond, T/Sgt Willfred S., 199

altitude, 43, 46, 48, 51-52, 59, 74, 103, 126-26, 129-30, 151-52, 154, 157-59, 173, 178, 181, 186-87

Aluminum Trail, 45, 52, 75

American Army Railway, 151, 166

anti-aircraft guns, 80

Approach Control Units, 172

Arnold, Gen. Hap, 28, 57, 61

Arthur Patterson, 105

ASC, 57, 124

Assam Valley, 30, 32, 39, 41-43, 49, 60, 74, 80, 87, 89, 103, 107, 114, 142, 210

ATC, 33, 57, 62, 68, 94, 124, 154, 159, 163

B-17, 126, 130

B-24, 68, 73,85, 89, 99, 103, 122, 135, 202,208

B-25s, 176, 179, 202

Baker, Col. Robert H., 160, 164, 166, 169, 174

Ballinger, Pfc. Don, 92, 95

Barksdale, Col. William S., 138, 143

Barrackapore, 153, 156, 161, 209

barracks, 85-86, 89-90

basha, 58, 73, 83, 86- 87, 89-91, 93, 96, 167

battle fatigue, 100, 104

Beckwith, Pfc. R. Vern, 106, 110

Bengal, 116, 120, 181

Bengal-Assam Railway, 58, 161, 166

Bengal Police, 119

Bissell, Maj. Gen. Clayton L., 95, 98

Black Area V, 112, 115-16,118, 142

Blacks, 115-16, 118, 147, 157, 207

Blue Box, 70

Blue, S/Sgt James H., 196

Boedecker, Captain 175

Boeing, 67

Bombay, 39

Bordene, Major, 58

Boxer Protocol, 27

Brahmaputra River, 39

British tents, 89

Brown, Joe E., 105

Brown, T/Sgt Roy T., 199

Burma Road, 24, 28, 29, 30, 33-37, 135

Bush, Lt. Henry C., 140, 145

Butler hangar, 139-40, 143-44, 148-49

C-1 09, 103, 123, 135-36, 146, 153, 169, 208

C-46 Curtiss Commando, 64, 73, 122-23, 125, 131-32, 135-36, 182, 208

C-47 Skytrain, 35, 40, 41, 45, 46, 59, 61, 63, 73, 82, 87, 123, 125, 132, 140, 149, 152, 156, 161, 169, 174, 178-81, 185, 190, 195, 202

C-54 Skymaster, 68, 117, 126, 132

C-87, 73, 121, 130-31, 145, 160-61

Cady, Lt. Harold, 118

Cailliez, Lt. Clovis C. V., 192-93, 197, 209-210, 212

Calcutta, 31, 34, 39, 107, 114, 159-60, 175, 200, 202-03, 209

Campbell, Jr., Col. George D., 169, 174

Case, F/O Emerald C., 198

CBI, 21, 23, 25, 29-32, 37, 39, 41-42, 51, 56-57, 62, 64-65, 67, 72, 75-76, 78, 80, 84, 92, 98, 100, 103, 105, 109-15, 127-28, 134, 142-43, 145-54, 162, 174, 177,

Chabua, 23, 39-41, 53, 56, 58-59, 74, 79, 82-83, 85-86, 89, 94, 102-03, 105, 109, 128, 135, 138, 142, 145, 149, 152, 154-55, 157-68, 170, 175, 178, 186-87, 194-96, 198, 205-06, 209, 214

Chanyi, 1155, 179-80, 195-96, 199

Chapel Club, 107

Chapman, Mr., 81-82

Charlie route,74

Chennault, Brig. Gen. Claire Lee, 28, 31, 37, 137, 168, 170, 178

Chiang Kai-shek, Generalissimo, 27-29, 31, 33, 35-36, 40, 98, 137, 142-43, 145-54, 162, 174, 177, 182-84, 188, 191, 202, 206, 214-18, 220

Chiang, Madam, 28, 37, 56, 137

Chihkiang, 179-80

China National Aviation Corporation (CNAC), 34-36, 60-61, 97

Chindit Forces, 36, 102, 156

Chindwin River Valley, 43
Chisholm, F/O Donald W. 194
Chowkidars, 186
Chungking, 33, 167, 176
Clague, Maj. Betty, 114
Clements, Jr., Maj. Vern, 115
Clements, F/O Warren W., 198
Cohen, Sammy, 105
Collen, S/Sgt. Robert B., 180
Collins, Capt. Louis, 204
Combat Cargo, 157, 216
Communist Party, 119
Congress Party, 119
Cooch Behar, 93
Corramore Tea Estate, 197
Courts Martial, 116-17

CRASHES, 52, 64, 126, 156, 184, 188, 194, 198-99

Cruze, Mrs. J. L., 105
Curtiss-Wright, 67, 126
Dead Reckoning, 58
DeLonza, 2/Lt Peter R., 198
Deragon, 89, 205

Dohbi, 91
dhobi ghat, 91
dhobi itch, 91
Dinjan, 59-60, 83, 156, 190, 208
Distinguished Flying Cross, 65, 106, 109
Distinguished Service Medal, 145
Donald, W. H., 29
Duke, Pfc Robert H., 199
Dunn, Capt. William P., 149
Easy Route 74
Eisenhower, Gen. Dwight, 141
Entertainment, 105
Falkenburg, Jinx, 105
Ferrying Command, 26, 98
Fisher, Herbert O., 126-27
Flight Officer 71, 103
Flight Officer Act, 71
Ford, Pfc Lee, 53
Fort Hertz, 56-57, 61, 85, 87, 99, 183
Foster, 2/Lt. Charles O., 78, 196
Foster, John M., 23, 56, 69-70, 73, 77-78, 140, 213
Franklin, Mr., 81-83

Gabharu River, 84

GALAHAD, 178-80

Galbraith, Maj. Charles O., 169

gangja, 118

Gillespie, F/O Max, 199

Giles, Maj. Gen. Barney M., 98

Godfrey, Brig. Gen. Stuart, 47

Greenwood Plantation, 81

Grevemberg, Maj., 87

GRUBWORM 177-78

Hancock, Sgt. John W., 198

Hardin, Brig. Gen./Col. Tom, 53, 53, 178

Harding, T/Sgt. Richard, 198

Harmon, Lt. Tom, 42

Harriman, Averell, 28, 33-34,

Hattialli Tea Estate, 81, 203

Himalayas, 30, 36, 42 49-50, 53, 55, 76, 131, 159, 178, 192, 215, 220

Hine, F/O Jason O., 199

Hoag, Brig. Gen. Earl, 53, 63, 178

Hog Bristles, 155

Holbrook, Roy, 28

Holms, Rev. Reuben, 107

Hump, 22-23, 25, 31, 34-38, 41, 43-45, 49-57, 60-64, 72-73, 75-77, 89, 91, 98, 103-04, 121-22, 124, 126-27, 129, 131, 134, 145, 154, 158, 164-65, 167, 169, 173, 176, 178, 183-85, 188-89, 195, 203-04, 207, 210, 212, 214-16, 220

Hump Express, 109

Ichigo, 176

Icing, 51-52, 54, 57, 85, 125-27, 153, 155, 167, 172, 175, 211

Irrawaddy, 30, 43

Jim Crow, 115

Johnson, Lonnie, 23, 43-44

Jorhat, 88, 90, 93, 103, 107-09, 113, 143, 145, 147-58, 174,197-99, 201, 206, 208

Jungle Indoctrination, 71

Kaplan, Maj. Morris, 202

Karachi, 39, 84-85, 156, 203, 206, 209, 210, 213

Kikujiro, Private Shumiro, 27

Kohima, 49, 211

Kumen Mountains, 43

Kunming, 30, 34-35, 45-46, 49, 52, 55, 58, 74, 87, 103, 155, 167, 169, 176-77, 179, 191, 194-96, 203, 208, 212

Kurmitola, 90, 152, 208

Kweiyang, 180, 196

L1, Piper, 184, 186

L-4s, Stinson, 184, 187

L-5s, 187

Lalmanir Hat, 60, 78, 93, 105 107, 111, 114, 118-19 143, 157, 161, 198, 200, 204, 208-09

Lambert, F/O Joseph M., 198

Lashio, 30

Ledo, 33-37, 86, 02, 118, 166, 169, 175, 187, 203, 210

Ledo Road, 35-37, 86, 118, 166, 187, 210

Lend-Lease, 26, 28, 30

Leonard, Gordon, 23, 103-04

Link Trainer, 70, 73

Lister Bags, 94

Long, F/O Maynard B., 196

LORAN, 758, 75

Loving, Jr., Robert M. (Pete), 23, 191-92

Loza, Sgt., Tony, 198

MacArthur, Gen., Douglas, 41

McBride, Sgt. George B., 193

McCoy, William, 23

McDonnell Douglas, 67

MacGregor, Major, 173

McLaren Jr., Capt C. F., 159

Manchuria, 27, 167

Mao Zedong, 27

Marco Polo Bridge, 27

Marshall Gen. George C., 41

Medevac, 102, 182, 202-04, 220

Mekong, 43

Mercury, 155

Merrill's Marauders, 37, 41, 56, 102, 178, 211, 218

Misamari, 23, 60, 73, 75, 81, 86-88, 91-92, 95, 108-09, 114-15, 117-19, 142, 145, 147, 149, 154, 157, 162-63, 173-74, 179, 190, 193-96, 199, 201, 205-06

Mishmis, 48

Mohanbari, 23, 41, 57-59, 81, 84-85, 107-08, 120, 124, 134, 149, 154-55, 159, 161, 172-73, 180, 183, 189-90, 197-99, 201, 210

Monsoon, 30, 41, 50-52, 108, 149, 158, 172, 200

Mooney, Pfc. Arville M., 196

Moore, Pvt. Clifford W., 194, 207

Morale, 85, 92-98, 100-01, 104, 109, 111, 113, 115-17, 120, 208

MOS, 87, 111-12, 114

Mountbatten, Lord, 32, 40, 137, 177

Mount Tali, 55

Muslim League, 119

Myitkyina, 30, 36, 74, 166-69, 172

Naga, 33, 48-49, 118, 198

Navigation, 42, 45, 58, 71, 74-75, 167, 170

Neibler, Harold W., 183

O'Brien, Pat, 105

O'Neal, 2/Lt. Andrew J., 199

Officer Candidate School (OCS), 61

Olds, Brig. Gen. Robert, 97

Operational Training Unit (OTU), 72-73

Oswalt, Walter R., 183

Paget, Mr., 81, 83

Pan American Airways, 23

Pathalpiam, 76

Patkai, 43

Patterson, Arthur, 105

Patton, Gen. George S., 215

Payne, Capt. John D., 58

Perry, Pvt. Herman, 118

Persian Gulf, 33

Phillips, Captain, 173

Pierce, Mr., 81

Pond, Jr., F/O Stephen B. 199

Pope Field, Fort Bragg, NC, 39, 84

Porter, Capt. John L "Blackie," 182-83

Pratt, Lt. Col. Claron, 115, 191

Major Donald C. Pricer, 184

Primary Pilot Training, 71

Prindle, 2/Lt. Donald, 199

Priorities and Traffic, 112, 173

Production Line Maintenance (PLM) 22, 79, 143, 145, 147

Project 7, 113

Project 8, 86-88

Proqhan, Mr., 197

Rachel, M/Sgt., 144

Radics, T/Sgt F.A., 199

Radio Operators, 75

Rangoon, 30-31

Reilly, Major, 87

Renshaw, Col. Harry, 76

Renshaw University, 109

rice paddies, 81

Rickenbacker, Capt. E. V., 98-100

Rierson, Pfc. James F., 180

Roberts, Jr., Capt. Robert C., 171, 179

Roberts, F/O Harry S., 196

Rockpile, 25, 43, 58

Rogers, Flt. Nurse Audrey, 204

ROOSTER Movement, 178-80

Royal Air Force (RAF), 78

Rupert's Beer, 104

Russian Turkestan Railroad, 33

Rust, Lt. Col. Gordon, 183

Salween, 43

Santung, 43

Schaffer, F/O Paul, 23, 103

Scott, 1/Lt. Vernon, 59

Sea of Bengal, 49

Seabolt, Maj. Robert J., 186

Search and Rescue, 22, 37, 64, 165, 182-83, 185, 190, 192, 197-98, 217-18, 220

Sergiopol, 33

Sevareid, Eric, 178

Sewell, Luke, 105

Shapiro, T/Sgt M. L., 198

Shillong, 77, 119, 211

Sikorsky R-4s, 184

Silk, 155

Singphos, 48

Smith, Gordon, 23, 144,

snack bar of Yunnanyi, 190

Sookerating, 41, 89, 155, 173 201, 205, 209

Soong, T. V., 28, 33-35

South East Asia Command, 32

Special Services, 107

Squadron E, 93

Stafford, Cpl. Frank E., 109

Stilwell, General Joseph, 31, 34, 41, 42, 98-99, 114, 175-76, 178, 203,

Stratemeyer, Major General George, 62, 98, 114, 166

Swisher, Capt. Keith D., 197

Sylhet, 58, 82, 190

Technical Orders (T.O.s), 133, 153, 161

Telecky, Capt. Frederick J., 193

Tenth Air Force, 97-98, 129, 162

Tezpur, 78, 89, 113, 174, 200-01, 205, 207-08

Thomas, Maj. Robert C. 78, 207

Thompson, 2/Lt. Kenneth E., 194

Thurman, Sgt. Robert, 109

Tibet, 57,

Tin, 155

TO&E, 94

Townsend, Major, 87

Tracy, 1/Lt. Harley L., 199

Tucker, Harry D., 183

Tungsten, 155

Tunner, Brig. General, William H., 21-22, 25, 43, 53, 63, 89, 100, 103, 111, 140, 145, 155-56, 188, 205, 211-12, 215-16

Turpin, F/O Howard, 198

Tushan, 180

U-64 Nordsman, 184, 187

Unger, Chaplin, 106

USO, 105

Vest, 2/Lt. L. E., 199

Vickers, 1/Lt. William F., 198

Vogel, 2/Lt. James H., 196

Volz, Maj. W. H., 191

vulture, 152

WACs, 114

Walker, 2/Lt. Charles, 172, 179

Walker, Dixie, 105

Walters, 1/Lt. Joe, 198

Warner, Paul "Pop," 105

Wedemeyer, General, 118

White, Col. Robert R., 148

Whitehurst, Capt., 191-92

Wilson, Maj. Robert M., 86

Wingate, General Orde C., 34, 56, 102

Wolframite, 155

Wood, Pvt. Robert A., 59

Wright, Maj. Robert, 48

Wynn, Keenan, 105

YOKE, 175-76

Yongdong River, 27

Yunnan Province, 32, 167, 176

Yunnanyi, 57

Zemerlin, Lt. John, 117

About the Author

Barbara F. Bates

B. F. Bates was born into the history of the Air Transport Command, listening to the stories of her father as he recounted his experiences flying the Hump. Growing up as an Air Force brat, Ms. Bates developed an understanding of the military and an appreciation for the sacrifices made by all the men and women who serve. With over thirty years of experience as an Instructional Designer for all military services except the US Coast Guard, Ms. Bates has developed training materials for five aircraft platforms and all of their aircrew members, as well as maintenance manuals.

She has a B.A. in History and English, and a Master's degree in Secondary Education. Her first book, Texoma Medical Center: The First Twenty-Five Years, was published in 1985. Ms. Bates lives in the Dallas suburbs with her son and family.

Franklin Publishing

The goal of Franklin Publishing is to enable Pastors, Evangelists, Missionaries, and Christian leaders and presenters to become published authors. Becoming a published author expands your influence and builds your ministry. You can write the book or sermon series which God has laid on your heart. We can walk that road with you.

www.FranklinPublishing.org

Come and visit our Facebook page and be sure to like and follow us to keep up with writing tips and new developments.

www.facebook.com/FranklinPublishing

www.ingramcontent.com/pod-product-compliance
Lightning Source LLC
Chambersburg PA
CBHW060500090426
42735CB00011B/2057